T0259986

Universitext

Editors (North America): S. Axler and K.A. Ribet

Aguilar/Gitler/Prieto: Algebraic Topology from a Homotopical Viewpoint
Aksoy/Khamsi: Nonstandard Methods in Fixed Point Theory
Andersson: Topics in Complex Analysis
Aupetit: A Primer on Spectral Theory
Bachman/Narici/Beckenstein: Fourier and Wavelet Analysis
Badescu: Algebraic Surfaces
Balakrishnan/Ranganathan: A Textbook of Graph Theory
Balser: Formal Power Series and Linear Systems of Meromorphic Ordinary Differential Equations
Bapat: Linear Algebra and Linear Models (2nd ed.)
Berberian: Fundamentals of Real Analysis
Blyth: Lattices and Ordered Algebraic Structures
Boltyanskii/Efremovich: Intuitive Combinatorial Topology (Shenitzer, trans.)
Booss/Bleecker: Topology and Analysis
Borkar: Probability Theory: An Advanced Course
Böttcher/Silbermann: Introduction to Large Truncated Toeplitz Matrices
Carleson/Gamelin: Complex Dynamics
Cecil: Lie Sphere Geometry: With Applications to Submanifolds
Chae: Lebesgue Integration (2nd ed.)
Charlap: Bieberbach Groups and Flat Manifolds
Chern: Complex Manifolds Without Potential Theory
Cohn: A Classical Invitation to Algebraic Numbers and Class Fields
Curtis: Abstract Linear Algebra
Curtis: Matrix Groups
Debarre: Higher-Dimensional Algebraic Geometry
Deitmar: A First Course in Harmonic Analysis (2nd ed.)
DiBenedetto: Degenerate Parabolic Equations
Dimca: Singularities and Topology of Hypersurfaces
Edwards: A Formal Background to Mathematics I a/b
Edwards: A Formal Background to Mathematics II a/b
Farenick: Algebras of Linear Transformations
Foulds: Graph Theory Applications
Friedman: Algebraic Surfaces and Holomorphic Vector Bundles
Fuhrmann: A Polynomial Approach to Linear Algebra
Gardiner: A First Course in Group Theory
Gårding/Tambour: Algebra for Computer Science
Goldblatt: Orthogonality and Spacetime Geometry
Gustafson/Rao: Numerical Range: The Field of Values of Linear Operators and Matrices
Hahn: Quadratic Algebras, Clifford Algebras, and Arithmetic Witt Groups
Heinonen: Lectures on Analysis on Metric Spaces
Holmgren: A First Course in Discrete Dynamical Systems
Howe/Tan: Non-Abelian Harmonic Analysis: Applications of $SL(2, R)$
Howes: Modern Analysis and Topology
Hsieh/Sibuya: Basic Theory of Ordinary Differential Equations
Humi/Miller: Second Course in Ordinary Differential Equations
Hurwitz/Kritikos: Lectures on Number Theory

(continued after index)

Volker Runde

A Taste of Topology

 Springer

Volker Runde
Department of Mathematical
    and Statistical Sciences
University of Alberta
Edmonton, Alberta
Canada T6G 2G1
vrunde@ualberta.ca

Mathematics Subject Classification (2000): 54–01, 55–01

Library of Congress Control Number: 2005924410

ISBN-10: 0-387-25790-X   Printed on acid-free paper
ISBN-13: 978-0-387-25790-7

Printed in the United States of America.

9 8 7 6 5 4 3 2 (Corrected Printing, 2007)

springer.com

Preface

If mathematics is a language, then taking a topology course at the undergraduate level is cramming vocabulary and memorizing irregular verbs: a necessary, but not always exciting exercise one has to go through before one can read great works of literature in the original language, whose beauty eventually—in retrospect—compensates for all the drudgery.

Set-theoretic topology leaves its mark on mathematics not so much through powerful theorems (even though there are some), but rather by providing a unified framework for many phenomena in a wide range of mathematical disciplines. An introductory course in topology is necessarily concept heavy; the nature of the subject demands it. If the instructor wants to flesh out the concepts with examples, one problem arises immediately in an undergraduate course: the students don't yet have a mathematical background broad enough that would enable them to understand "natural" examples, such as those from analysis or geometry. Most examples in such a course therefore tend to be of the concocted kind: constructions, sometimes rather intricate, that serve no purpose other than to show that property XY is stronger than property YX whereas the converse is false. There is the very real danger that students come out of a topology course believing that freely juggling with definitions and contrived examples is what mathematics—or at least topology—is all about.

The present book grew out of lecture notes for Math 447 (Elementary Topology) at the University of Alberta, a fourth-year undergraduate course I taught in the winter term 2004. I had originally planned to use [SIMMONS 63] as a text, mainly because it was the book from which I learned the material. Since there were some topics I wanted to cover, but that were not treated in [SIMMONS 63], I started typing my own notes and making them available on the Web, and in the end I wound up writing my own book. My audience included second-year undergraduates as well as graduate students, so their mathematical background was inevitably very varied. This fact has greatly influenced the exposition, in particular the selection of examples. I have made an effort to present examples that are, firstly, not self-serving and, secondly, accessible for students who have a background in calculus and elementary algebra, but not necessarily in real or complex analysis.

It is clear that an introductory topology text only allows for a limited degree of novelty. Most topics covered in this book can be found in any other book on the subject. I have thus tried my best to make the presentation as fresh and accessible as possible, but whether I have succeeded depends very much on my readers' tastes. Besides, in a few points, this books treats its material differently than—to my knowledge, at least—any other text on the subject.

- Baire's theorem is derived from Bourbaki's Mittag-Leffler theorem;
- Nets are extensively used, and, in particular, we give a fairly intuitive proof—using nets—of Tychonoff's theorem due to Paul R. Chernoff [CHERNOFF 92];
- The complex Stone–Weierstraß theorem is obtained via Silvio Machado's approach [MACHADO 77] with a particularly short and elegant proof due to Thomas J. Ransford [RANSFORD 84].

With a given syllabus and a limited amount of classroom time, every instructor in every course has to make choices on what to cover and what to omit. These choices will invariably reflect his or her own tastes and biases, in particular, when it comes to omissions. The topics most ostensibly omitted from this book are: filters and uniform spaces. I simply find nets, with all the parallels between them and sequences, far more intuitive than filters when it comes to discussing convergence (others may disagree). Treating uniform spaces in an introductory course is a problem, in my opinion, due to the lack of elementary, yet natural, examples that aren't metric spaces in the first place.

Any book, even if there is only one author named on the cover, is to some extent an accomplishment of several people. This one is no exception, and I would like to thank Eva Maria Krause for her thorough and insightful proofreading of the entire manuscript. Of course, without my students—their feedback and enthusiasm—this book would not have been written. I hope that taking the course was as much fun for them as teaching it was for me, and that they had *A Taste of Topology* that will make their appetite for mathematics grow in the years to come.

Volker Runde
Edmonton, 2007

Contents

Preface v

Introduction 1

1   Set Theory 5
     1.1   Sets and Functions 5
     1.2   Cardinals 13
     1.3   Cartesian Products 17
     Remarks 21

2   Metric Spaces 23
     2.1   Definitions and Examples 23
     2.2   Open and Closed Sets 28
     2.3   Convergence and Continuity 34
     2.4   Completeness 40
     2.5   Compactness for Metric Spaces 52
     Remarks 59

3   Set-Theoretic Topology 61
     3.1   Topological Spaces—Definitions and Examples 61
     3.2   Continuity and Convergence of Nets 72
     3.3   Compactness 79
     3.4   Connectedness 89
     3.5   Separation Properties 100
     Remarks 107

4   Systems of Continuous Functions 109
     4.1   Urysohn's Lemma and Applications 109
     4.2   The Stone–Čech Compactification 116
     4.3   The Stone–Weierstraß Theorems 121
     Remarks 129

5   **Basic Algebraic Topology** 133
      5.1   Homotopy and the Fundamental Group 133
      5.2   Covering Spaces 148
      Remarks 154

A   **The Classical Mittag-Leffler Theorem
      Derived from Bourbaki's** 157

B   **Failure of the Heine–Borel Theorem
      in Infinite-Dimensional Spaces** 161

C   **The Arzelà–Ascoli Theorem** 165

References 169

Index 171

List of Symbols

(0), 68
$\| \cdot \|$, 24
$\| \cdot \|_1$, 24
$\| \cdot \|_\infty$, 24
$\bigcap \{S : S \in \mathcal{S}\}$, 8
$\bigcup \{S : S \in \mathcal{S}\}$, 8
$\in$, 5
∞, 34
$\notin$, 5
∂S, 33
$\prod \{S : S \in \mathcal{S}\}$, 18
$\prod_{i \in \mathbb{I}} S_i$, 18
$\sim$, 134
$\simeq$, 136
$\subset$, 6
$\subsetneq$, 6
$\varnothing$, 5
2^κ, 16

$\aleph_0$, 16
(a, b), 6
$[a, b]$, 6
$(a, b]$, 6
$[a, b)$, 6
$A \cap B$, 8
$A \cup B$, 8
$A \setminus B$, 8
$A_{r,R}[x_0]$, 135

βX, 118
B_n, 143
$B_r(x_0)$, 28
$B_r[x_0]$, 30
$B(S, Y)$, 24
$\mathcal{B}_x$, 65

$\mathfrak{c}$, 16
$\mathbb{C}$, 5
$\mathbb{C}_\infty$, 86
$C([0, 1])$, 24
$C(X, Y)$, 42

$C_b(X, Y)$, 42
$C_0(X, \mathbb{F})$, 126
cl, 67

d, 24
diam, 44
dim, 40
dist, 34
dist_F, 122

ϕ_α, 144
$f|_A$, 10
$f(A)$, 10
$f^{-1}(B)$, 10
$f \circ g$, 11
$\mathbb{F}$, 24
f^{-1}, 12
f_*, 142
$f : S \to T$, 10
$F(S, Y)$, 65

$[\gamma]$, 141
$\gamma_1 \odot \gamma_2$, 98
γ^{-1}, 98

$H(\Omega)$, 159

id_S, 10

$\lim_\alpha x_\alpha$, 74
$\lim_{n \to \infty} x_n$, 35
$L(\mathcal{U})$, 58

μ, 95

$\mathbb{N}$, 5
$\mathbb{N}_0$, 5
$\mathcal{N}_x$, 29
$N_{f,C,\epsilon}$, 65
$\mathfrak{N}_x$, 64

π, 8

$\pi_1(X, x_0)$, 138
$\pi_n(X, x_0)$, 155
$\mathfrak{p}$, 63
$\mathfrak{P}(S)$, 7
$P(X, x_0)$, 138
$P(X; x_0, x_1)$, 138

$\mathbb{Q}$, 5

$\mathbb{R}$, 5
$R(f; \mathcal{P}, \xi)$, 74

$\overline{S}$, 30
$|S| = |T|$, 13
$|S| \geq |T|$, 15
$|S| > |T|$, 15
$\overset{\circ}{S}$, 34
$|S| \leq |T|$, 15
$|S| < |T|$, 15
$\mathrm{Spec}(R)$, 63
$\mathbb{S}^{n-1}$, 90
S^2, 9
$S \times T$, 9
$S^{\mathbb{I}}$, 18
S^n, 17

$\mathcal{T}$, 61
$\mathcal{T}_C$, 65
$\mathcal{T}_\infty$, 86

$V(I)$, 63

χ_n, 95
$(x_\alpha)_\alpha$, 74
$(x_\alpha)_{\alpha \in \mathbb{A}}$, 74
$x_\alpha \to x$, 74
(X, d), 24
X_∞, 86
$(x_n)_{n=1}^\infty$, 10
$(x_n)_{n=m}^\infty$, 10
$x_n \to x$, 35
$x \preceq y$, 19
$(X, \mathcal{T})$, 62
$\left(\left(\tilde{X}, \tilde{\mathcal{T}} \right), p \right)$, 149
(x, y), 9

Y_x, 94

$\mathbb{Z}$, 5

Introduction

The present book is an introduction to set-theoretic topology (and to a tiny little bit of algebraic topology).

The prerequisites for a reader who wants to read this book profitably are modest. First of all, a basic familiarity with set-theoretic terminology is necessary. It is also helpful to have a good background in calculus (both in one variable and in several variables), not so much because we rely on results from calculus, but rather because having been exposed to a certain concept (continuity, for instance) in the relatively concrete framework of calculus will make it easier to grasp the same concept in the more abstract and less intuitive setting of general topology. For some examples and exercises, as well as for the last two chapters, some familiarity with the definitions of basic algebraic objects—rings, ideals, groups, and so on—is also needed.

Chapter One gives a quick introduction to the set theory required for the remaining four chapters. Since the reader is presumed to have encountered basic set-theoretic notions (set, element, subset, etc.) before, we decided to keep it brief. Based on a naive notion of set, we introduce the basic set-theoretic constructions, such as unions and intersections, define functions, and discuss cardinalities. We use Zorn's lemma to derive the axiom of choice. This chapter is somewhat less rigorous than the remaining ones of this book: we never outline a system of formal axioms for set theory. As the main purpose of this book is to serve as an introduction to topology, a version of this chapter that would have been up to the same standards of rigor as the rest of the book would simply have taken too much space.

Chapters Two to Four deal with set-theoretic topology. Roughly speaking, set theoretic topology is about providing a conceptual framework to meaningfully speak about continuity: equip sets with just enough structure so that it makes sense to say whether maps between them are continuous. The generality of the concepts and results of set-theoretic topology makes its basics indispensable for anyone who wishes to study any branch of analysis or geometry in some depth.

Chapter Two introduces metric spaces, discusses topological concepts in the metric context, and treats continuity and the peculiar features of compactness in the metric space situation. There are many notions and results contained in this chapter that are not actually about metric spaces, but about topological spaces. Hence, some material from Chapter Two is duplicated in Chapter Three. From a pedagogical point of view, it is probably better to treat (some) topological concepts first in the relatively concrete setting of metric spaces, than do it in full generality right away. Whenever a result in Chapter Two holds true for general topological spaces, a proof is given that is as topological as possible, that is, without direct reference to metrics, so that later, when the result becomes available in its full generality, a simple reference to the proof in the metric case is sufficient. Baire's theorem is obtained in a somewhat unusual way, namely as an application of Bourbaki's Mittag-Leffler theorem.

General topological spaces are introduced in Chapter Three. We define topological spaces by axiomatizing the notion of an open set, but alternative approaches—through neighborhoods or a closure operation—are also covered. We then proceed to the definition of continuity: since sequences turn out to be inadequate tools for the study of topological spaces, we first give a definition of continuity that avoids any notion of convergence. Subsequently, we introduce nets and use them to characterize continuity and various topological phenomena. Due to the formal parallels between nets and sequences, this allows an approach to general topological spaces that still formally resembles the treatment of the metric case. In particular, the use of nets allows us to give a relatively simple proof of Tychonoff's theorem due to Paul R. Chernoff [CHERNOFF 92]. We discuss connectedness and path connectedness, as well as their local variants. The chapter closes with an overview of separation properties, from T_0 to normality.

In Chapter Four, we turn to the actual *raison d'être* of topological spaces: the study of continuous functions (here, with values in $\mathbb{R}$ or $\mathbb{C}$). Urysohn's lemma along with its consequences—Urysohn's metrization theorem and Tietze's extension theorem—are presented, and subsequently the Stone–Čech compactification of a completely regular space is introduced. The chapter ends with a discussion of the real and complex Stone–Weierstraß theorems, both on compact and locally compact spaces; the proof is based on Silvio Machado's approach [MACHADO 77] in the short and elegant form given by Thomas J. Ransford [RANSFORD 84].

Even though they are both called topology, set-theoretic and algebraic topology have relatively little in common. They share the objects of study, topological spaces, and before one can start learning algebraic topology, one needs a certain familiarity with set theoretic topology, but surprisingly little is needed: most of algebraic topology can do perfectly well without Tychonoff's theorem, separation axioms, and Urysohn's lemma with all its consequences. Algebraic topology is about studying algebraic invariants of topological spaces: to a given topological space, an algebraic object (often a group) is assigned in such a way that if the spaces can be identified, then so can the associated algebraic objects. Since the tools of algebra are generally very powerful, this can be used to tell that two spaces are different because the associated algebraic invariants can be told apart.

In Chapter Five, we take a brief look at one of those invariants: the fundamental group. We introduce the notions of homotopy and path homotopy, and define the fundamental group of a topological space at a given base point. We compute the fundamental group for convex subsets of normed spaces (it is trivial) and for the unit circle in $\mathbb{R}^2$ (it is $\mathbb{Z}$). Since the fundamental groups of homotopically equivalent spaces are isomorphic, we conclude that the closed unit disc and the unit circle—or more generally, any closed annulus—in $\mathbb{R}^2$ cannot be homotopically equivalent (let alone homeomorphic). In order to identify the fundamental group of the unit circle as $\mathbb{Z}$, we take an even briefer look at the concept of a covering space. We show that paths in a topological space can be lifted to a covering space in such a way that path homotopies are preserved.

Each section of each chapter ends with exercises, which (what else?) are intended to help deepen the reader's understanding of the material. Within each section, exercises are just referred to by their numbers; from other sections, references to a particular exercise are made by combining the section number and the exercise number. For example, Exercise 4 in Section 3.2 is referred to as Exercise 4 throughout Section 3.2, but as Exercise 3.2.4 from anywhere else.

Each chapter has an unnumbered remarks section at its end. These sections contain remarks of an historical nature, views to beyond the actual contents of the chapter, and suggestions for further reading.

There are three appendices. Their contents could have been fitted into the five chapters of the book (Appendix A into Section 2.4, Appendix B into Section 2.5, and Appendix C into Section 3.3). The material in all three appendices, however, is more analytical than topological in nature, and Appendix A also requires some knowledge of the theory of holomorphic functions from the reader.

1

Set Theory

If an introduction to topology is about learning some essential vocabulary of the language of mathematics, then set theory provides the alphabet in which this vocabulary is expressed.

1.1 Sets and Functions

Since the main focus of this book is topology and not set theory, we adopt a completely naive attitude towards sets.

"Definition" 1.1.1. *A* set *is a collection of certain objects considered as a whole.*

This is, of course, far from being a precise definition (that's why the word is in quotation marks): What is a "collection"? What are "certain objects"? And what does it mean to consider a collection of certain objects—whatever that may be—"as a whole"? Instead of dwelling on these questions (and becoming overly formalistic), we content ourselves with fleshing out the notion of a set with some examples:

Example 1.1.2. The collection of positive integers (excluding 0) is a set denoted by $\mathbb{N}$. Also the nonnegative integers (including 0), the integers, the rational numbers, the reals, and the complex numbers constitute sets that are denoted by $\mathbb{N}_0$, $\mathbb{Z}$, $\mathbb{Q}$, $\mathbb{R}$, and $\mathbb{C}$, respectively.

We also want to call a collection of nothing a set: this is the *empty set* denoted by $\varnothing$.

If x is one of the objects collected in the set S, we call x an *element* of S and denote this by $x \in S$ (we then say that x "is contained" in S or "lies in" S); if x is not an element of S, we write $x \notin S$.

Example 1.1.3. We have $\sqrt{2} \in \mathbb{R}$, but $\sqrt{2} \notin \mathbb{Q}$.

If T and S are sets, then T is called a *subset* of S (in symbols: $T \subset S$) if each element of T is also an element of S (with some risk of ambiguity, we then also say that T "is contained in" S).

Examples 1.1.4. (a) We have

$$\mathbb{N} \subset \mathbb{N}_0 \subset \mathbb{Z} \subset \mathbb{Q} \subset \mathbb{R} \subset \mathbb{C}.$$

(b) Since $\varnothing$ has no elements, it is a subset of every other set.

If $T \subset S$ and $S \subset T$, we say that the two sets S and T are equal and write $S = T$. In the case $T \subset S$, but $S \neq T$, we use the symbol $T \subsetneq S$; we then call the subset T of S *proper*.

Example 1.1.5. Clearly,

$$\mathbb{N} \subsetneq \mathbb{N}_0 \subsetneq \mathbb{Z} \subsetneq \mathbb{Q} \subsetneq \mathbb{R} \subsetneq \mathbb{C}$$

holds.

Let S be a set, and let P be any property that is either satisfied by a particular element of S or isn't. Then

$$\{x \in S : x \text{ satisfies } P\}$$

is the collection of all elements of S satisfying P and is a subset of S.

Examples 1.1.6. (a) The even numbers

$$\{x \in \mathbb{Z} : 2 \text{ divides } x\}$$

form a subset of $\mathbb{Z}$.

(b) Let $a, b \in \mathbb{R} \cup \{-\infty, \infty\}$ with $a \leq b$. Then the *open interval*

$$(a, b) := \{x \in \mathbb{R} : a < x < b\},$$

the *closed interval*

$$[a, b] := \{x \in \mathbb{R} : a \leq x \leq b\},$$

as well as the *half-open intervals*

$$(a, b] := \{x \in \mathbb{R} : a < x \leq b\} \quad \text{and} \quad [a, b) := \{x \in \mathbb{R} : a \leq x < b\}$$

are subsets of $\mathbb{R}$. Note that we allow $a = b$. Hence, $(a, a) = (a, a] = [a, a) = \varnothing$ and $[a, a]$, which is $\varnothing$ if a is $-\infty$ or ∞ and consists of the one element a if $a \in \mathbb{R}$, are also intervals; we call such intervals *degenerate*.

Since sets themselves are also "certain objects," a collection of sets should again be a set.

Example 1.1.7. If S is any set, then its *power set* $\mathfrak{P}(S)$ is defined as the collection of all subsets of S. For example, $S \in \mathfrak{P}(S)$ and $\varnothing \in \mathfrak{P}(S)$.

Given finitely many distinct objects $x_1, \ldots, x_n$, we denote by $\{x_1, \ldots, x_n\}$ the set made up by them. Sets arising in this fashion are called *finite*, and we say that n is the *cardinality* of the set S or that S has n elements. Sets of cardinality one (i.e., consisting of one single element) are sometimes referred to as *singletons*. The way the elements $x_1, \ldots, x_n$ are ordered doesn't affect the set $\{x_1, \ldots, x_n\}$ at all: for instance, $\{1, 2, 3\} = \{2, 1, 3\} = \{3, 1, 2\} = \cdots$.

Proposition 1.1.8. *Let S be a set having n elements. Then $\mathfrak{P}(S)$ has cardinality 2^n.*

Proof. If $n = 0$, then $S = \varnothing$, so that $\mathfrak{P}(S) = \{\varnothing\}$; that is, $\mathfrak{P}(S)$ has $1 = 2^0$ elements.

Suppose that the claim holds for $n \in \mathbb{N}_0$, and let S have $n + 1$ elements; that is, $S = \{x_1, \ldots, x_n, x_{n+1}\}$. Let $T := \{x_1, \ldots, x_n\}$. Any subset of S thus must either be a subset of T or contain x_{n+1}. Therefore we have:

number of subsets of S

$\quad$ = number of subsets of T + number of subsets of S containing x_{n+1}.

By the induction hypothesis, there are 2^n subsets of T. Given any subset A of S containing x_{n+1}, we may define a subset $A' := \{x \in A : x \neq x_{n+1}\}$ of T. Clearly, each such subset A of S yields a unique subset A' of T. Moreover, whenever B is a subset of T, we can define a unique subset $\tilde{B}$ of S by letting $\tilde{B} := \{x \in S : x \in B \text{ or } x = x_{n+1}\}$. It is clear that $\widetilde{(A')} = A$ for each subset A of S containing x_{n+1} and that $\left(\tilde{B}\right)' = B$ for each subset B of T. Hence, there are as many subsets of S containing x_{n+1} as there are subsets of T. Again, by the induction hypothesis, we obtain that there are 2^n subsets of S containing x_{n+1}.

Eventually, we have

$$\text{number of subsets of } S = 2^n + 2^n = 2 \cdot 2^n = 2^{n+1},$$

as claimed. $\square$

Forming sets out of sets again, however, can be dangerous.

Example 1.1.9 (Russell's antinomy). Since collections of sets are sets again, the collection of all sets should again be a set. Given any set S, it either contains itself as an element or it doesn't. The property of a set to contain itself as an element looks strange, all examples of sets one naively comes up with don't have it, but that's beside the point: it is a legitimate property of sets, which they may or may not have. Hence, we can form the subset

$$\mathcal{S} := \{S : S \text{ is a set not containing itself as an element}\}$$

of the set of all sets. Does the set S contain itself as an element? If so, then (by its own definition!) it should not be contained in itself, which is nonsense. On the other hand, if S is not contained in itself, then its definition again forces the contrary to be true. This doesn't make sense at all.

What goes wrong in Example 1.1.9? Apparently, we cannot just form arbitrary collections of objects and label them sets. Roughly speaking, the collection of all sets is simply too "large" (whatever that may mean precisely) to be a set again. We thus have to impose restrictions. Since this is a book on elementary topology and not on set theory, we avoid trouble with Russell's antinomy the easy way: all sets we encounter are supposed to be subsets of one very large set, the *universe*, which is large enough for us to do everything we need (e.g., form power sets) in order to do topology, but too small for monsters like the "set of all sets."

We now give our first formal definition.

Definition 1.1.10. *Let S be a set, and let $A, B \subset S$. Then:*

(a) *The* union *$A \cup B$ of A and B is the set consisting of all elements of S that are contained in A or in B.*

(b) *The* intersection *$A \cap B$ of A and B is the set consisting of all elements of S that are contained both in A and in B. If $A \cap B = \varnothing$, we say that A and B are* disjoint.

(c) *The* set-theoretic difference *$A \setminus B$ of A and B is the set consisting of all elements of S contained in A, but not in B. We call $S \setminus A$ the* complement *of A in S.*

Examples 1.1.11. (a) We have, for example,

$$(-2, \pi) \cup (\pi, 7] = (-2, 7] \setminus \{\pi\}.$$

(b) Let A consist of the prime numbers, and let B be the set of all even numbers. Then $A \cap B = \{2\}$ holds.

(c) The set $\mathbb{R} \setminus \mathbb{Q}$ consists of all irrational numbers.

(d) For any set S and $A, B \subset S$, we have $(A \setminus B) \cap B = \varnothing$.

Definition 1.1.10(a) and (b) easily extend to arbitrary families of sets. Given a collection $\mathcal{S}$ of sets (all subsets of one given set), its *union* is

$$\bigcup \{S : S \in \mathcal{S}\} := \{x : x \text{ is contained in one of the sets } S \in \mathcal{S}\}$$

and its *intersection* is

$$\bigcap \{S : S \in \mathcal{S}\} := \{x : x \text{ is contained in all of the sets } S \in \mathcal{S}\}.$$

Often, the sets in our collection $\mathcal{S}$ will be *indexed* by some other set, $\mathbb{I}$ say, which is then called an *index set*. Informally, this means that each $S \in \mathcal{S}$ gets an *index* $i \in \mathbb{I}$ attached to it as some sort of tag, so that it can be identified as S_i. For $\mathcal{S}$, we then write $(S_i)_{i\in \mathrm{I}}$ or simply $(S_i)_i$ if no confusion can arise about I. In this situation, we write $\bigcup_{i\in \mathrm{I}} S_i$ and $\bigcap_{i\in \mathrm{I}} S_i$ (or simply, $\bigcup_i S_i$ and $\bigcap_i S_i$ if no confusion can arise) instead of $\bigcup\{S : S \in \mathcal{S}\}$ and $\bigcap\{S : S \in \mathcal{S}\}$, respectively.

Examples 1.1.12. (a) We have

$$\bigcup_{n=1}^{\infty}[-n,n] = \mathbb{R} \quad \text{and} \quad \bigcap_{n=1}^{\infty}\left[-\frac{1}{n},\frac{1}{n}\right] = \{0\}.$$

(b) Any set S is the union of its singleton subsets:

$$S = \bigcup_{x\in S}\{x\}.$$

(c) For $n \in \mathbb{N}$, let

$$S_n := \mathbb{Q} \cap \left[\sqrt{2}-\frac{1}{n},\sqrt{2}+\frac{1}{n}\right].$$

Since $\sqrt{2} \notin \mathbb{Q}$, it follows that $\bigcap_{n=1}^{\infty} S_n = \varnothing$.

Euclidean 2-space $\mathbb{R}^2$ is the 2-dimensional plane of geometric intuition. Each point in the plane can be identified through a pair (x,y) with $x,y \in \mathbb{R}$, where x is the first and y the second coordinate of the point. The ordered pair (x,y) must not be confused with the set $\{x,y\}$: we have $\{1,2\} = \{2,1\}$, but $(1,2) \neq (2,1)$.

In more formal terms, we define the following.

Definition 1.1.13. *Let S and T be sets. Then the* Cartesian product *of S and T is defined as*

$$S \times T := \{\{x,\{x,y\}\} : x \in S,\ y \in T\}.$$

For $x \in S$ and $y \in T$, we denote $\{x,\{x,y\}\} \in S \times T$ by (x,y) and call it the ordered pair *with first coordinate x and second coordinate y. We write S^2 instead of $S \times S$.*

Our next definition is also an attempt to formalize an already familiar notion.

Definition 1.1.14. *Let S and T be sets. A* function *(or* map*) f from S to T is a subset of $S \times T$ with the following properties.*

(a) *For each $x \in S$, there is $y \in T$ such that $(x,y) \in f$;*
(b) *Whenever $(x,y_1),(x,y_2) \in f$ holds, we have $y_1 = y_2$.*

The set S is called the domain *of f.*

This definition looks worlds apart from the intuitive notion of a function as something that assigns values in its range to the points in its domain. In fact, it isn't: it is just a more precise wording of that intuitive notion. Given $x \in S$, we have, by Definition 1.1.14(a), a value $y \in T$ such that $(x, y) \in f$, which is uniquely determined by Definition 1.1.14(b). We may thus denote that particular y by $f(x)$ and say that "f maps x to $f(x)$". To indicate that f is a function from the set S to the set T, we write $f : S \to T$, and for $(x, y) \in f$, we use the notation $y = f(x)$. The expression

$$f : S \to T, \quad x \mapsto f(x)$$

then stands for $\{(x, f(x)) : x \in S\} \subset S \times T$.

Examples 1.1.15. (a) Let S and T be sets. Then

$$S \times T \to S, \quad (x, y) \mapsto x$$

is the *coordinate projection* onto S. Similarly, the coordinate projection onto T is defined.
(b) Let S be any set. Then $\{(x, x) \in S^2 : x \in S\}$ is the *identity map*

$$\mathrm{id}_S : S \to S, \quad x \mapsto x$$

on S. (If no confusion can arise about the set S, we also simply write id.)
(c) If S and T are sets, $f : S \to T$ is a function, and $A \subset S$, then the *restriction* of f to A is defined as

$$f|_A := \{(x, f(x)) : x \in A\}.$$

Clearly, $f|_A : A \to T$ is again a function.
(d) Let S be any set. A map from $\mathbb{N}$ to S is called a *sequence* in S; instead of $x : \mathbb{N} \to S$, we then often write $(x_n)_{n=1}^{\infty}$. If the domain of x is not $\mathbb{N}$ but a subset of $\mathbb{N}_0$ of the form $\{n : n \geq m\}$ for some $m \in \mathbb{N}_0$, we still speak of a sequence and denote it by $(x_n)_{n=m}^{\infty}$. We call a sequence $(y_k)_{k=1}^{\infty}$ a *subsequence* of $(x_n)_{n=1}^{\infty}$ if there are $n_1 < n_2 < \cdots$ in $\mathbb{N}$ such that $y_k = x_{n_k}$ for $k \in \mathbb{N}$.

The following definitions are useful throughout.

Definition 1.1.16. *Let S and T be sets, let $f : S \to T$ be a function, and let $A \subset S$ and $B \subset T$.*

(a) *The* image *$f(A)$ of A under f is the subset $\{f(x) : x \in A\}$ of T. The image of S under f is also called the* range *of f.*
(b) *The* inverse image *$f^{-1}(B)$ of B under f is the subset $\{x \in S : f(x) \in B\}$ of S.*

Examples 1.1.17. (a) Let

$$f \colon \mathbb{R} \to \mathbb{R}, \quad x \mapsto x^2.$$

Then $f(\mathbb{R}) = [0, \infty)$, $f([2, 3]) = [4, 9]$, $f^{-1}(\mathbb{R}) = f^{-1}([0, \infty)) = \mathbb{R}$, $f^{-1}((-\infty, 2)) = f^{-1}([0, 2)) = (-\sqrt{2}, \sqrt{2})$, $f^{-1}([-\pi, -e]) = \varnothing$, and so on, hold.

(b) Let

$$f \colon \mathbb{R}^2 \to \mathbb{R}, \quad (x, y) \mapsto x - y.$$

Then, for example,

$$f^{-1}(\{7\}) = \{(x + 7, x) : x \in \mathbb{R}\}$$

and

$$f([-3, 7] \times [13, 42)) = (-45, -6]$$

hold.

Definition 1.1.18. *Let S and T be sets, and let $f \colon S \to T$ be a function. Then:*

(a) *f is called* injective *(or an* injection*) if $f(x_1) \neq f(x_2)$ for all $x_1, x_2 \in S$ with $x_1 \neq x_2$.*

(b) *f is called* surjective *(or a* surjection*) if, for each $y \in T$, there is $x \in S$ with $f(x) = y$.*

(c) *f is called* bijective *(or a* bijection*) if it is both injective and surjective.*

Whether a function is injective or surjective or bijective (or none of them) depends, of course, on the sets S and T.

Example 1.1.19. The function

$$f \colon S \to T, \quad x \mapsto x^2$$

is

- Bijective if $S = T = [0, \infty)$,
- Injective, but not surjective if $S = [0, \infty)$ and $T = \mathbb{R}$,
- Surjective, but not injective if $S = \mathbb{R}$ and $T = [0, \infty)$, and
- Neither injective nor surjective if $S = T = \mathbb{R}$.

The bijective maps are especially important for us.

Definition 1.1.20. *Let R, S, and T be sets, and let $g \colon R \to S$ and $f \colon S \to T$ be functions. Then the* composition *$f \circ g$ of f and g is the function*

$$f \circ g \colon R \to T, \quad x \mapsto f(g(x)).$$

Proposition 1.1.21. *Let S and T be sets. Then the following are equivalent for a function $f \colon S \to T$.*

(i) f *is bijective.*

(ii) *There is a function* $g\colon T \to S$ *such that* $f \circ g = \mathrm{id}_T$ *and* $g \circ f = \mathrm{id}_S$.

In this case, the function g in (ii) *is unique and called the* inverse function *of f (denoted by f^{-1}).*

Proof. (i) $\implies$ (ii): Define $g\colon T \to S$ as follows. Given $y \in T$, the surjectivity of f yields $x \in S$ with $f(x) = y$. The injectivity of f ascertains that x is unique. Hence, we may define $g(y) := x$. From this definition, it is clear that $f(g(y)) = y$ and that $g(f(x)) = x$.

(ii) $\implies$ (i): Let $y \in T$. Letting $x := g(y)$, we obtain $x \in S$ with $f(x) = y$. Hence, f is surjective. If $x_1, x_2 \in S$ are such that $f(x_1) = f(x_2)$, we obtain that $x_1 = g(f(x_1)) = g(f(x_2)) = x_2$, so that f is also injective.

Since the function g in (ii) has to assign, to each $y \in T$, the unique $x \in S$ with $f(x) = y$, it is clear that g is unique. $\square$

Examples 1.1.22. (a) The function

$$f\colon [0, \infty) \to [0, \infty), \quad x \mapsto x^2$$

is bijective and its inverse function is

$$f^{-1}\colon [0, \infty) \to [0, \infty), \quad x \mapsto \sqrt{x}.$$

(b) The tangent function is injective when restricted to $\left(-\frac{\pi}{2}, \frac{\pi}{2}\right)$. Since the image of that interval under tan is all of $\mathbb{R}$, we obtain an inverse function—arctan—of tan from $\mathbb{R}$ to $\left(-\frac{\pi}{2}, \frac{\pi}{2}\right)$.

In view of Definition 1.1.16, one might have second thoughts whether it is a good idea to denote the inverse function of a bijective map f by f^{-1}. What is $f^{-1}(B)$ supposed to mean? The inverse image of B under f or the image of B under f^{-1}. As it turns out, the symbol $f^{-1}(B)$ means the same in both contexts (Exercise 5 below).

Exercises

1. Prove *de Morgan's rules*: for a family $(S_i)_{i \in I}$ of subsets of a given set S, the identities

$$S \setminus \left(\bigcup_{i \in I} S_i \right) = \bigcap_{i \in I} (S \setminus S_i) \quad \text{and} \quad S \setminus \left(\bigcap_{i \in I} S_i \right) = \bigcup_{i \in I} (S \setminus S_i)$$

hold.

2. Let S and T be sets, and let $f\colon S \to T$ be a map. Show that f is injective if and only if $f|_A$ is injective for each subset A of S containing at most two elements.

3. Let S and T be sets, and let $f\colon S \to T$ be a function. Show that
   (a) f is injective if and only if $f^{-1}(f(A)) = A$ for all subsets A of S.
   (b) f is surjective if and only if $f(f^{-1}(B)) = B$ for all subsets B of T, which is the case if and only if $f(S) = T$.

4. Let f be a bijective map with inverse function f^{-1}. Show that f^{-1} is also bijective.

5. Let S and T be sets, let $f\colon S \to T$ be bijective, and let $B \subset T$. Show that the image of B under f^{-1} is the inverse image of B under f.

6. Let R, S, and T be sets, and let $g\colon R \to S$ and $f\colon S \to T$ be functions. Show that
   (a) if both f and g are injective, then so is $f \circ g$.
   (b) if both f and g are surjective, then so is $f \circ g$.
   (c) if both f and g are bijective, then so is $f \circ g$.
   How, if f and g are both bijective, is $(f \circ g)^{-1}$ expressed in terms of f^{-1} and g^{-1}?

7. Let S and T be sets, and let $f\colon S \to T$ be a function. Show that:
   (a) f is injective if and only if there is a function $g\colon T \to S$ such that $g \circ f = \mathrm{id}_S$.
   (b) f is surjective if there is a function $g\colon T \to S$ such that $f \circ g = \mathrm{id}_T$.

1.2 Cardinals

Which of the sets $\{1,2,3\}$ and $\{1,2,3,4\}$ is larger? The second one, of course: it contains the first one as a proper subset. What if neither of two sets is contained in the other one; for example, what about $\{1,2,3\}$ and $\{\clubsuit, \diamondsuit, \heartsuit, \spadesuit\}$? Of course, $\{\clubsuit, \diamondsuit, \heartsuit, \spadesuit\}$ is larger than $\{1,2,3\}$: it has the same number of elements as $\{1,2,3,4\}$, which we know to be larger than $\{1,2,3\}$. All in all, it is intuitively clear, for finite sets, what it means when we say that one of them is larger than the other or that two of them have the same size. But what does it mean if we make such a statement about sets that aren't finite, that is, about *infinite* sets?

Since $\mathbb{N}$ is a proper subset of $\mathbb{N}_0$, one might think that $\mathbb{N}_0$ is "larger" than $\mathbb{N}$. On the other hand, the map

$$\mathbb{N}_0 \to \mathbb{N}, \quad n \mapsto n+1$$

is easily seen to be bijective, so that each element of $\mathbb{N}_0$ corresponds to precisely one element of $\mathbb{N}$. Hence, $\mathbb{N}_0$ and $\mathbb{N}$ should have "the same number of elements" and thus be of equal size. This second approach turns out to be the appropriate one when it comes to dealing with "sizes" of infinite sets.

Definition 1.2.1. *Two sets S and T are said to have* the same cardinality, *in symbols:* $|S| = |T|$, *if there is a bijective function $f\colon S \to T$.*

Examples 1.2.2. (a) Two finite sets have the same cardinality if and only if they have the same number of elements. In particular, a subset T of a finite set S has the same cardinality as S if and only if $S = T$.

(b) If $|R| = |S|$ and $|S| = |T|$, then $|R| = |T|$ holds.

(c) The sets $\mathbb{N}$ and $\mathbb{N}_0$ have the same cardinality, even though $\mathbb{N}$ is a proper subset of $\mathbb{N}_0$.

(d) For each $x \in \mathbb{R}$, let $\lfloor x \rfloor \in \mathbb{Z}$ denote the largest integer less than or equal to x. The map

$$\mathbb{N} \to \mathbb{Z}, \quad n \mapsto (-1)^n \left\lfloor \frac{n}{2} \right\rfloor$$

is bijective, so that $|\mathbb{N}| = |\mathbb{Z}|$.

The last example shows that strange things happen when one starts dealing with cardinalities of infinite sets. (We show below that even $\mathbb{Q}$ has the same cardinality as $\mathbb{N}$.) Do all infinite sets have the same cardinality? This is not true.

Theorem 1.2.3. *There is no surjective map from $\mathbb{N}$ onto $(0,1)$.*

Proof. We require a fundamental fact from analysis.

Every number $r \in (0,1)$ has a decimal expansion $r = 0.\sigma_1\sigma_2\sigma_3\ldots$; that is, $r = \sum_{k=1}^{\infty} \frac{\sigma_k}{10^k}$, with $\sigma_k \in \{0,1,\ldots,9\}$ for $k \in \mathbb{N}$.

Assume towards a contradiction that there is a surjective map $r : \mathbb{N} \to (0,1)$. For each $n \in \mathbb{N}$, the number $r(n)$ has a decimal expansion:

$$r(n) = 0.\sigma_1(n)\,\sigma_2(n)\,\sigma_3(n)\ldots.$$

We then define $r \in (0,1)$ by giving its decimal expansion $r = 0.\sigma_1\sigma_2\sigma_3\ldots$:

$$\sigma_k = \begin{cases} 2, & \text{if } \sigma_k(k) \neq 2, \\ 3, & \text{if } \sigma_k(k) = 2. \end{cases}$$

Since $(0,1) = r(\mathbb{N})$, there must be $n \in \mathbb{N}$ with $r = r(n)$ and thus $\sigma_n = \sigma_n(n)$, which is absurd by the definition of r. $\square$

Instead of 2 and 3 in the proof of Theorem 1.2.3, we could have used any other two digits from $\{0,1,\ldots,8\}$ (but excluding 9).

Corollary 1.2.4. *The sets $\mathbb{N}$ and $(0,1)$ do not have the same cardinality.*

Hence, $\mathbb{N}$ and $(0,1)$ represent "different sizes" of infinity, so that (c) in the following definition actually covers some ground.

Definition 1.2.5. *A set S is called*

(a) Countably infinite *if it has the same cardinality as $\mathbb{N}$,*
(b) Countable *if it is finite or countably infinite, and*
(c) Uncountable *if it is infinite, but not countable.*

If a set is countably infinite, each of its elements corresponds uniquely to a positive integer. We therefore sometimes denote countable (possibly finite) sets as $\{x_1, x_2, x_3, \ldots\}$, with the understanding that the enumeration $x_1, x_2, \ldots$ breaks off after some point if the set is finite.

Having defined what it means for arbitrary sets to have the same cardinality, we now turn to defining what it means for a set to have a cardinality less than another set.

Definition 1.2.6. *Let S and T be sets. We say that the* cardinality *of S is less than or equal to the cardinality of T, in symbols: $|S| \leq |T|$ or $|T| \geq |S|$, if there is an injective map from S to T. If $|S| \leq |T|$, but not $|S| = |T|$, we write $|S| < |T|$ or $|T| > |S|$.*

If S and T are finite sets with $|S| \leq |T|$ and $|S| \geq |T|$, it is immediate that $|S| = |T|$. Somewhat surprisingly, this remains true for arbitrary sets.

Theorem 1.2.7 (Cantor–Bernstein). *Let S and T be sets such that $|S| \leq |T|$ and $|S| \geq |T|$. Then $|S| = |T|$ holds.*

Proof. Let $f : S \to T$ and $g : T \to S$ be injective maps. Even though f and g need not be bijective, both maps become bijective if considered as maps $f : S \to f(S)$ and $g : T \to g(T)$, so that it makes sense to speak of $f^{-1}: f(S) \to S$ and $g^{-1}: g(T) \to T$.

We call an element $x \in S$ an ancestor of itself of degree zero. If $x \in g(T)$, we call $g^{-1}(x)$ an ancestor of x of degree one. If $g^{-1}(x) \in f(S)$, we call $f^{-1}(g^{-1}(x))$ an ancestor of x of degree two, and if $f^{-1}(g^{-1}(x)) \in g(T)$, we say that $g^{-1}(f^{-1}(g^{-1}(x)))$ is an ancestor of x of degree three. This pattern can go on indefinitely or it breaks off at some point. Anyhow, we can define

$$\deg(x) := \sup\{n \in \mathbb{N}_0 : x \text{ has an ancestor of degree } n\} \in \mathbb{N}_0 \cup \{\infty\}.$$

Let

$$S_\infty := \{x \in S : \deg(x) = \infty\}, \qquad S_{\text{even}} := \{x \in S : \deg(x) \in \mathbb{N}_0 \text{ is even}\},$$

and

$$S_{\text{odd}} := \{x \in S : \deg(x) \in \mathbb{N}_0 \text{ is odd}\}.$$

Clearly, each element of x lies in precisely one of the sets S_∞, S_{even}, or S_{odd}. With an analogous argument, we obtain a similar partition T_∞, T_{even}, and T_{odd} of T.

The following are then easily verified: $f(S_\infty) = T_\infty$, $f(S_{\text{even}}) = T_{\text{odd}}$, and $g^{-1}(S_{\text{odd}}) = T_{\text{even}}$. We can thus define $h : S \to T$ by letting

$$h(x) := \begin{cases} f(x), & \text{if } x \in S_\infty \cup S_{\text{even}}, \\ g^{-1}(x), & \text{if } x \in S_{\text{odd}}. \end{cases}$$

We claim that h is bijective.

To see that h is surjective, let $y \in T$. We need to show that there is $x \in S$ with $h(x) = y$.

Case 1: $y \in T_\infty$. Since $f(S_\infty) = T_\infty$, there is $x \in S_\infty$ with $f(x) = y$. From the definition of h, it follows that $h(x) = f(x) = y$.

Case 2: $y \in T_{\text{even}}$. Since $g^{-1}(S_{\text{odd}}) = T_{\text{even}}$, there is $x \in S_{\text{odd}}$ such that $h(x) = g^{-1}(x) = y$.

Case 3: $y \in T_{\text{odd}}$. Since $f(S_{\text{even}}) = T_{\text{odd}}$, there is $x \in S_{\text{even}}$ such that $h(x) = f(x) = y$.

To prove the injectivity of h, let $x_1, x_2 \in S$ be such that $h(x_1) = h(x_2)$. Since $h(S_\infty) = T_\infty$, $h(S_{\text{even}}) = T_{\text{odd}}$, and $h(S_{\text{odd}}) = T_{\text{even}}$, and since T_∞, T_{even}, T_{odd} are mutually disjoint, we conclude that either $x_1, x_2 \in S_\infty$ or $x_1, x_2 \in S_{\text{even}}$ or $x_1, x_2 \in S_{\text{odd}}$. Since f and $g^{-1}|_{S_{\text{odd}}}$ are injective, it follows from the definition of h that $x_1 = x_2$. $\square$

Examples 1.2.8. (a) Let S and T be countable sets. It is easy to see that $S \times T$ is again countable if S or T is finite. We therefore suppose that S and T are both countably infinite. Let $f : \mathbb{N} \to S$ and $g : \mathbb{N} \to T$ be bijective. Fixing $y \in T$,
$$\mathbb{N} \to S \times T, \quad n \mapsto (f(n), y)$$
is an injective map. On the other hand,
$$S \times T \to \mathbb{N}, \quad (x, y) \mapsto 2^{f^{-1}(x)} 3^{g^{-1}(y)}$$
is also injective (due to the uniqueness of the prime factorization in $\mathbb{N}$). The Cantor–Bernstein theorem thus yields that $S \times T$ is also countably infinite.

(b) The map
$$f : \mathbb{Z} \times \mathbb{N} \mapsto \mathbb{Q}, \quad (n, m) \mapsto \frac{n}{m}$$
is surjective. By the previous example, $|\mathbb{Z} \times \mathbb{N}| = |\mathbb{N}|$ holds, i.e., there is a bijective map $g : \mathbb{N} \to \mathbb{Z} \times \mathbb{N}$. Then $f \circ g : \mathbb{N} \to \mathbb{Q}$ is surjective, so that $|\mathbb{Q}| \le |\mathbb{N}|$ by Exercise 1 below. Since trivially $|\mathbb{N}| \le |\mathbb{Q}|$, it follows that $\mathbb{Q}$ is countably infinite.

(c) The function
$$\mathbb{R} \to \mathbb{R}, \quad t \mapsto \frac{1}{\pi}\left(\arctan t + \frac{\pi}{2}\right)$$
maps $\mathbb{R}$ bijectively onto $(0, 1)$. Hence, $\mathbb{R}$ and $(0, 1)$ have the same cardinality. From the previous example and Theorem 1.2.3, it follows that $|\mathbb{R}| > |\mathbb{Q}|$.

Two finite sets have the same cardinality if and only if they have the same number of elements: given a finite set, the class—not a set!—of all sets that can be mapped bijectively onto it is represented by one positive integer, namely the number of elements of the given set. In analogy, for any set, the class of all sets with the same cardinality as that set is defined as a *cardinal number* or simply *cardinal*. The positive integers then are nothing but particular cardinals; since they are represented by finite sets, we call them *finite cardinals*; all other cardinals are called *infinite*. Usually, cardinals are denoted by letters from the middle of the Greek alphabet, such as κ or λ. The cardinality of $\mathbb{N}$ is commonly denoted by $\aleph_0$ ($\aleph$, spelled aleph, is the first letter of the Hebrew alphabet) whereas $\mathfrak{c}$ (for continuum) stands for $|\mathbb{R}|$. If κ is any cardinal, represented by a set S, then the cardinality of its power set is often denoted by 2^κ (which makes sense in view of Proposition 1.1.8).

From Theorem 1.2.3 and Exercise 2 below, it is clear that both $\aleph_0 < \mathfrak{c}$ and $\aleph_0 < 2^{\aleph_0}$. But more is true.

Proposition 1.2.9. $\mathfrak{c} = 2^{\aleph_0}$.

Proof. Given $S \subset \mathbb{N}$, define $(\sigma_n(S))_{n=1}^\infty$ by letting $\sigma_n(S) = 1$ if $n \in S$ and $\sigma_n = 2$ if $n \notin S$, and let $r(S) := \sum_{n=1}^\infty \frac{\sigma_n(S)}{10^n}$. Then

$$\mathfrak{P}(\mathbb{N}) \to (0,1), \quad S \mapsto r(S)$$

is injective, so that $2^{\aleph_0} \leq \mathfrak{c}$.

For the converse inequality, we use the fact that every $r \in (0,1)$ not only has a decimal expansion, but also a binary one: $r = \sum_{n=1}^\infty \frac{\sigma_n(r)}{2^n}$ with $\sigma_n(r) \in \{0,1\}$ for $n \in \mathbb{N}$. Hence, every number in $(0,1)$ can be represented by a string of zeros and ones. This representation, however, is not unique: for example, both $1000\ldots$ and $0111\ldots$ represent the number $\frac{1}{2}$. This, however, is the only way ambiguity can occur. Hence, whenever $r \in (0,1)$ has a period $\bar{1}$, we convene to pick its nonperiodic binary expansion. In this fashion, we assign, to each $r \in (0,1)$, a unique sequence $(\sigma_n(r))_{n=1}^\infty$ in $\{0,1\}$. The map

$$(0,1) \to \mathfrak{P}(\mathbb{N}), \quad r \mapsto \{n \in \mathbb{N} : \sigma_n(r) = 1\}$$

is then injective, so that $2^{\aleph_0} \geq \mathfrak{c}$. $\square$

Exercises

1. Let S be a set, and let $f \colon \mathbb{N} \to S$ be a surjective map. Show that there is a map $g \colon S \to \mathbb{N}$ such that $f \circ g = \mathrm{id}_S$, and conclude that $|S| \leq |\mathbb{N}|$.
2. Let S be a set. Show that $|S| < |\mathfrak{P}(S)|$. (*Hint:* Assume that there is a surjective map $f \colon S \to \mathfrak{P}(S)$, and consider the set $\{x \in S : x \notin f(x)\}$.)
3. Let $(S_n)_{n=1}^\infty$ be a sequence of countable sets. Show that $\bigcup_{n=1}^\infty S_n$ is countable.
4. A real number is called *algebraic* if there is a nonzero polynomial p with rational coefficients such that $p(x) = 0$ and *transcendental* otherwise (for instance, $\sqrt{2}$ is algebraic, but π is transcendental). Show that the set of algebraic numbers is countable, and conclude that there are uncountably many transcendental numbers.
5. Show that $\aleph_0 \leq \kappa$ for each infinite cardinal κ.
6. Let κ be a cardinal such that $\kappa < \aleph_0$. Show that κ is finite.

1.3 Cartesian Products

In Definition 1.1.13, we formally defined the Cartesian product of two sets. It is easy to extend Definition 1.1.13 to Cartesian products of finitely many sets, $S_1, \ldots, S_n$ say: simply define

$$S_1 \times S_2 \times \cdots \times S_n := (S_1 \times \cdots \times S_{n-1}) \times S_n$$

through induction on n (if $S_1 = \cdots = S_n =: S$, we often write S^n). The elements of $S_1 \times \cdots \times S_n$ are then ordered n-tuples $(x_1, \ldots, x_n)$ with *coordinates* $x_j \in S_j$, which are inductively defined as

$$(x_1, x_2, \ldots, x_n) := ((x_1, \ldots, x_{n-1}), x_n) \qquad (x_j \in S_j,\ j = 1, \ldots, n).$$

Now, let $\mathcal{S}$ be an *arbitrary* (i.e., possibly infinite) collection of sets. How should their Cartesian product $\prod\{S : S \in \mathcal{S}\}$ be defined? To answer this question, we first take a closer look at Cartesian products of finitely many sets.

Example 1.3.1. Let $S_1, \ldots, S_n$ be sets. Since it is a standing hypothesis of ours that all sets we encounter are subsets of one giant universe, we may form the union $S_1 \cup \cdots \cup S_n$. Let $f \colon \{1, \ldots, n\} \to S_1 \cup \cdots \cup S_n$ be a function such that $f(j) \in S_j$ for $j = 1, \ldots, n$. Then $(f(1), \ldots, f(n))$ is an element of $S_1 \times \cdots \times S_n$. Conversely, if $(x_1, \ldots, x_n) \in S_1 \times \cdots \times S_n$, the function

$$f \colon \{1, \ldots, n\} \to S_1 \cup \cdots \cup S_n, \quad j \mapsto x_j$$

satisfies $f(j) \in S_j$ for $j = 1, \ldots, n$. Hence, another way to describe $S_1 \times \cdots \times S_n$ is as the set of all functions f from $\{1, \ldots, n\}$ to $S_1 \cup \cdots \cup S_n$ such that $f(j) \in S_j$ for $j = 1, \ldots, n$.

This motivates the following definition.

Definition 1.3.2. *Let $\mathcal{S}$ be a nonempty collection of sets. Then the* Cartesian product $\prod\{S : S \in \mathcal{S}\}$ *is defined to be the collection of all functions $f \colon \mathcal{S} \to \bigcup\{S : S \in \mathcal{S}\}$ such that $f(S) \in S$ for all $S \in \mathcal{S}$.*

If the sets in $\mathcal{S}$ are indexed, by $\mathbb{I}$ say, so that $\mathcal{S} = (S_i)_{i \in \mathbb{I}}$, we also write $\prod_{i \in \mathbb{I}} S_i$ (and $\prod_i S_i$ if no confusion can arise) for their Cartesian product. If $S_i = S$ holds for some set S and for all $i \in \mathbb{I}$, we also write $S^{\mathbb{I}}$ instead of $\prod_{i \in \mathbb{I}} S$, which is nothing but the set of all functions from $\mathbb{I}$ to S.

It is straightforward that $\prod\{S : S \in \mathcal{S}\} \neq \varnothing$ whenever $\mathcal{S}$ is finite and $S \neq \varnothing$ for each $S \in \mathcal{S}$. (The same can be shown if $\mathcal{S}$ is countable; see Exercise 2 below.) The question of whether $\prod\{S : S \in \mathcal{S}\} \neq \varnothing$ for an *arbitrary* nonempty collection $\mathcal{S}$ of nonempty sets, however, is surprisingly intricate. At first glance, the answer seems to be straightforward: we need to find a *choice function*, a function $f \colon \mathcal{S} \to \bigcup\{S : S \in \mathcal{S}\}$ that chooses an element $f(S)$ from each set $S \in \mathcal{S}$. If $\mathcal{S}$ is finite, say $\mathcal{S} = \{S_1, \ldots, S_n\}$, this is easy: pick $f(S_1) \in S_1$, which is possible because $S_1 \neq \varnothing$, then choose $f(S_2) \in S_2$, and continue in this fashion until $f(S_n) \in S_n$ has been chosen. If $\mathcal{S}$ is arbitrary, there is no procedure like this that would allow us to find a choice function. Nevertheless, it still seems plausible that $\prod\{S : S \in \mathcal{S}\}$ is always nonempty; there is an element in every $S \in \mathcal{S}$ and therefore there should be a way of choosing one element from every such S. To prove this statement, however, requires more powerful set-theoretic tools than we have seen so far.

By a *relation* $\mathcal{R}$ on a set S, we mean a subset of S^2.

Definition 1.3.3. *A relation $\mathcal{R}$ on a set S is called an* ordering *if the following are satisfied.*

(a) $(x, x) \in \mathcal{R}$ *for each $x \in S$;*

(b) If $(x, y), (y, z) \in \mathcal{R}$, then $(x, z) \in \mathcal{R}$;

(c) If $(x, y), (y, x) \in \mathcal{R}$, then $x = y$.

Instead of $(x, y) \in \mathcal{R}$, we rather write $x \preceq y$, and we use the symbol $\preceq$ to denote the ordering $\mathcal{R}$. A set S equipped with an ordering $\preceq$ is called *ordered*; if, for any $x, y \in S$, one of $x \preceq y$ or $y \preceq x$ holds, we say that S is *totally ordered*.

Examples 1.3.4. (a) The real numbers with their usual order are a totally ordered set.

(b) Let S be any set. For $A, B \subset S$ define

$$A \preceq B \quad :\Longleftrightarrow \quad A \subset B.$$

This turns $\mathfrak{P}(S)$ into an ordered set that is not totally ordered if S has more than one element.

Definition 1.3.5. *Let S be an ordered set.*

(a) *An element $x \in S$ is called an* upper bound *for $T \subset S$ if $y \preceq x$ for all $y \in T$.*

(b) *An element $x \in S$ is called* maximal *if there is no $y \in S$, $x \neq y$, such that $x \preceq y$.*

We can now formulate *Zorn's lemma*.

Axiom 1.3.6 (Zorn's lemma). *Let S be an ordered, nonempty set with the property that each nonempty, totally ordered subset of S has an upper bound. Then S has maximal elements.*

The label "lemma" for Zorn's lemma is highly deceptive: it is not a lemma, but an axiom of set theory that cannot be proven without other—equally nontrivial—hypotheses.

It is fair to say that the statement of Zorn's lemma is far from being intuitive. With its help, however, the following can be proven:

Theorem 1.3.7. *Let $(S_i)_{i \in \mathbb{I}}$ be a nonempty family of nonempty sets. Then $\prod_{i \in \mathbb{I}} S_i$ is nonempty.*

Proof. Let $\mathcal{P}$ be the collection of all pairs $(\mathbb{J}_f, f)$, where $\varnothing \neq \mathbb{J} \subset \mathbb{I}$ and $f : \mathbb{J}_f \to \bigcup_{j \in \mathbb{J}_f} S_j$ is such that $f(j) \in S_j$ for all $j \in \mathbb{J}_f$. Clearly, $\mathcal{P}$ is not empty: fix $i \in \mathbb{I}$, let $x \in S_i$, and define $f : \{i\} \to S_i$ by letting $f(i) = x$; it is clear that $(\{i\}, f) \in \mathcal{P}$.

We define an order on $\mathcal{P}$ by letting, for $(\mathbb{J}_f, f), (\mathbb{J}_g, g) \in \mathcal{P}$,

$$(\mathbb{J}_f, f) \preceq (\mathbb{J}_g, g) \quad :\Longleftrightarrow \quad \mathbb{J}_f \subset \mathbb{J}_g \text{ and } g|_{\mathbb{J}_f} = f.$$

Let $\mathcal{Q}$ be a nonempty, totally ordered subset of $\mathcal{P}$. Let

$$\mathbb{J}_g := \bigcup \{\mathbb{J}_f : (\mathbb{J}_f, f) \in \mathcal{Q}\},$$

and define $g: \mathbb{J}_g \to \bigcup_{j \in \mathbb{J}_g} S_j$ as follows: for $j \in \mathbb{J}_g$, there is $(\mathbb{J}_f, f) \in \mathcal{Q}$ such that $j \in \mathbb{J}_f$; set $g(j) = f(j)$. Since $\mathcal{Q}$ is *totally* ordered, it is easily seen that g is well defined; that is, the value of $g(j)$ does not depend on the particular choice of $(\mathbb{J}_f, f) \in \mathcal{Q}$ with $j \in \mathbb{J}_f$. It is clear that $(\mathbb{J}_g, g) \in \mathcal{P}$ is an upper bound for $\mathcal{Q}$.

Zorn's lemma then yields a maximal element $(\mathbb{J}_{\max}, f_{\max})$ of $\mathcal{P}$. Assume that $\mathbb{J}_{\max} \neq \mathbb{I}$ (i.e., there is $i_0 \in \mathbb{I} \setminus \mathbb{J}_{\max}$). Fix $x_0 \in S_{i_0}$, and define $\tilde{f}: \mathbb{J}_{\max} \cup \{i_0\} \to \bigcup_{j \in \mathbb{J}_{\max}} S_j \cup S_{i_0}$ by letting

$$\tilde{f}(j) := \begin{cases} f_{\max}(j), & j \in \mathbb{J}_{\max}, \\ x_0, & j = i_0. \end{cases}$$

It follows that $\left(\mathbb{J}_{\max} \cup \{i_0\}, \tilde{f}\right) \in \mathcal{P}$ with $(\mathbb{J}_{\max}, f_{\max}) \preceq \left(\mathbb{J}_{\max} \cup \{i_0\}, \tilde{f}\right)$, but $(\mathbb{J}_{\max}, f_{\max}) \neq \left(\mathbb{J}_{\max} \cup \{i_0\}, \tilde{f}\right)$, which contradicts the maximality of $(\mathbb{J}_{\max}, f_{\max})$. Hence, $\mathbb{J}_{\max} = \mathbb{I}$ holds, so that $f_{\max} \in \prod_i S_i$. $\square$

On the surface, the statement of Theorem 1.3.7 seems to be far more plausible than that of Zorn's lemma, so that one is tempted to ask if one *really* needs Zorn's lemma to prove Theorem 1.3.7. One does. In fact, one can suppose that the assertion of Theorem 1.3.7—which is then called the *axiom of choice*—is true and then deduce Zorn's lemma from it.

Exercises

1. Let $S \neq \varnothing$ be a set. A relation $\mathcal{R}$ on S is called an *equivalence relation* if it is
   (i)   *Reflexive* (i.e., $(x, x) \in \mathcal{R}$ for each $x \in S$),
   (ii) *Symmetric* (i.e., if $(x, y) \in \mathcal{R}$, then $(y, x) \in \mathcal{R}$ for all $x, y \in S$), and
   (iii) *Transitive* (i.e., if $x, y, z \in S$ are such that $(x, y), (y, z) \in \mathcal{R}$, then $(x, z) \in \mathcal{R}$ holds).
   (Often, one writes, $x \sim y$, $x \approx y$, etc., instead of $(x, y) \in \mathcal{R}$.) Given $x \in S$, the *equivalence class* of x (with respect to a given equivalence relation $\mathcal{R}$) is defined to consist of those $y \in S$ for which $(x, y) \in \mathcal{R}$. Show that two equivalence classes are either disjoint or identical.
2. Let $(S_n)_{n=1}^\infty$ be a sequence of nonempty sets. Show *without invoking Zorn's lemma* that $\prod_{n=1}^\infty S_n$ is not empty.
3. A *Hamel basis* of a (possibly infinite-dimensional) vector space (over an arbitrary field) is a linearly independent subset whose linear span is the whole space. Use Zorn's lemma to show that every nonzero vector space has a Hamel basis.
4. Let R be a commutative ring with identity 1. An ideal $\mathfrak{m}$ is called maximal if $\mathfrak{m} \subsetneq R$ and $I = \mathfrak{m}$ or $I = R$ for each ideal I of R with $\mathfrak{m} \subset I \subset R$. Use Zorn's lemma to show that every proper ideal in R (i.e., one that doesn't equal R) is contained in a maximal ideal.
5. Use Theorem 1.3.7 to prove a converse of Exercise 1.1.7(b): For sets S and T and a surjective function $f: S \to T$, there is a function $g: T \to S$ such that $f \circ g = \mathrm{id}_T$.
6. Let S and T be sets. Show that $|S| \leq |T|$ if and only if there is a surjective map $f: T \to S$.

Remarks

Expressing mathematics in set-theoretic terms seems so completely natural nowadays that it is hard to imagine to do mathematics any way otherwise. Nevertheless, set theory didn't enter the mathematical stage before the second half of the nineteenth century, and when it did, not everyone greeted it with applause.

Set theory is the brainchild of Georg Cantor. Born in the Russian capital St. Petersburg in 1845, to a German father and a Russian mother, he studied mathematics in Germany and Switzerland and obtained his doctorate for a thesis on number theory from Berlin in 1867. From number theory, he moved to analysis, and investigations into the convergence of Fourier series led him to eventually develop set theory. By the early 1870s, Cantor had proven that the algebraic numbers were countable whereas the reals weren't. From the late 1870s to the mid 1880s, he systematically laid down the foundations of set theory in a series of papers.

Cantor's approach to set theory was "naive" in the sense that it used the intuitive, but ultimately insufficiently rigorous "Definition" 1.1.1 of a set. Later in his life, Cantor himself discovered the first disturbing paradoxes in his intellectual constructions.

Russell's antinomy, from 1901, is named after its discoverer, the English mathematician, philosopher, Nobel laureate (for literature), and political activist, Bertrand Russell.

The contradictions in Cantor's set theory were eventually overcome with the help of rigorous axiomatizations that impose restrictions on how sets could be formed from other sets, but still allow enough freedom for everyday mathematical work. The system of axioms most commonly used today by set theorists is called *Zermelo–Fraenkel set theory* (and sometimes *Zermelo–Fraenkel–Skolem set theory*), named after its creators and abbreviated as ZF. The vast majority of mathematicians today are working within the framework of ZF, even though most of them would probably flunk a quiz on what precisely its axioms are. For a very accessible introduction to ZF-style axiomatic set theory see [HALMOS 74]; despite its title, the set theory presented there is not naive in any way.

By the early twentieth century, set theory had become accepted by most mathematicians. David Hilbert worded it memorably: "No one shall expel us from the paradise that Cantor has created for us."

The axiom of choice (AC) is independent of ZF: both ZF + AC, that is, ZF with the axiom of choice added as an additional axiom, and ZF + ¬AC, where the negation of AC is added as an axiom, are free from contradictions. The axiom of choice and Zorn's lemma (ZL) are equivalent in the sense that precisely the same theorems can be proven in ZF + AC and ZF + ZL. A third statement equivalent to the axiom of choice and Zorn's lemma, respectively, is the *well-ordering principle*. A *well ordering* on a set S is a total order such that each non-empty subset of S has a minimal element; the canonical order on $\mathbb{N}$, for example, is a well-ordering. The well-ordering principle asserts:

There is a well ordering on every nonempty set.

For proofs of how to derive AC, ZL, and the well-ordering principle from one another, see [HALMOS 74].

Most mathematicians working today accept the axiom of choice/Zorn's lemma/the well-ordering principle for pragmatic reasons: it enables them to prove useful theorems in their respective fields. One of the most important theorems in set-theoretic topology, for instance, is Tychonoff's theorem (Theorem 3.3.21 below); Zorn's lemma is indispensable for its proof. It is probably not exaggerated to say that most of functional analysis and abstract algebra would collapse without the axiom of choice.

As Cantor proved, $\aleph_0 < \mathfrak{c}$ holds, and he himself already asked if there was any cardinal strictly between $\aleph_0$ and $\mathfrak{c}$. The belief that no such cardinal exists is called the *continuum hypothesis*. His failure to prove it troubled Cantor deeply. In 1900, David Hilbert gave a famous speech at the International Congress of Mathematicians in Paris, in which he identified twenty-three open problems as central to mathematical research in the coming century; among them was the question of whether the continuum hypothesis was true. This problem was solved, in a certain sense, by the American Paul Cohen more than half a century later. The continuum hypothesis relates to ZF + AC in the same way as AC does to ZF: they are independent. Cohen received the Fields medal for his discovery in 1966.

From his late thirties onward, Cantor suffered from bouts of depression. The sometimes bitter controversies surrounding his mathematical ideas didn't help. Due to his depression, he was hospitalized several times throughout his later years; he was in a sanatorium when he died from a heart attack in 1918.

2

Metric Spaces

What is the minimum of structure one needs to have on a set in order to be able to speak of continuity?

If f is a function defined on a subset of $\mathbb{R}$—or, more generally, of Euclidean n-space $\mathbb{R}^n$—we say that f is continuous at x_0 if "$f(x)$ approaches $f(x_0)$ as x approaches x_0." With ϵ and δ, this statement can be made sufficiently precise for mathematical purposes.

> *For each $\epsilon > 0$, there is $\delta > 0$ such that $|f(x) - f(x_0)| < \epsilon$ for all x such that $|x - x_0| < \delta$.*

Crucial for the definition of continuity thus seems to be that we can measure the distance between two real numbers (or, rather, two vectors in Euclidean n-space).

If we want to speak of continuity of functions defined on more general sets, we should thus have a meaningful way to speak of the distance between two points of such a set: this, in a nutshell, is the idea behind a metric space.

2.1 Definitions and Examples

In Euclidean 2-space, the distance between two points (x_1, x_2) and (y_1, y_2) is defined as $\sqrt{(x_1 - y_1)^2 + (x_2 - y_2)^2}$. More generally, in Euclidean n-space $\mathbb{R}^n$, one defines, for $x = (x_1, \ldots, x_n)$ and $y = (y_1, \ldots, y_n)$, their distance as

$$d(x, y) := \sqrt{\sum_{j=1}^{n} (x_j - y_j)^2}.$$

The Euclidean distance has the following properties.

1. $d(x, y) \geq 0$ for all $x, y \in \mathbb{R}^n$ with $d(x, y) = 0$ if and only if $x = y$;
2. $d(x, y) = d(y, x)$ for all $x, y \in \mathbb{R}^n$;
3. $d(x, z) \leq d(x, y) + d(y, z)$ for $x, y, z \in \mathbb{R}^n$.

In the definition of a metric space, these three properties of the Euclidean distance are axiomatized.

Definition 2.1.1. *Let X be a set. A* metric *on X is a map $d \colon X \times X \to \mathbb{R}$ with the following properties:*

(a) $d(x, y) \geq 0$ *for all $x, y \in X$ with $d(x, y) = 0$ if and only if $x = y$* (positive definiteness);

(b) $d(x, y) = d(y, x)$ *for all $x, y \in X$* (symmetry);

(c) $d(x, z) \leq d(x, y) + d(y, z)$ *for $x, y, z \in X$* (triangle inequality).

A set together with a metric is called a metric space.

We often denote a metric space X whose metric is d by (X, d); sometimes, if the metric is obvious or irrelevant, we may also simply write X.

Examples 2.1.2. (a) $\mathbb{R}^n$ with the Euclidean distance is a metric space.

(b) Let (X, d) be a metric space, and let Y be a subset of X. Then the restriction of d to $Y \times Y$ turns Y into a metric space of its own. The metric space $(Y, d|_{Y \times Y})$ is called a *subspace* of X. In particular, any subset of $\mathbb{R}^n$ equipped with the Euclidean distance is a subspace of $\mathbb{R}^n$.

(c) Let E be a linear space (over $\mathbb{F} = \mathbb{R}$ or $\mathbb{F} = \mathbb{C}$). A *norm* on E is a map $\| \cdot \| \colon E \to \mathbb{R}$ such that: (i) $\|x\| \geq 0$ for all $x \in E$ with $\|x\| = 0$ if and only if $x = 0$; (ii) $\|\lambda x\| = |\lambda| \|x\|$ for $\lambda \in \mathbb{F}$ and $x \in E$; (iii) $\|x + y\| \leq \|x\| + \|y\|$ for all $x, y \in E$ (a linear space equipped with a norm is called a *normed space*). For $x, y \in E$, define

$$d(x, y) := \|x - y\|.$$

This turns E into a metric space. For example, let E be $C([0, 1], \mathbb{F})$, the space of all continuous $\mathbb{F}$-valued functions on $[0, 1]$. Then there are several norms on E, for example, $\| \cdot \|_1$ defined by

$$\|f\|_1 := \int_0^1 |f(t)| \, dt \qquad (f \in E)$$

or $\| \cdot \|_\infty$ given by

$$\|f\|_\infty := \sup\{|f(t)| : t \in [0, 1]\} \qquad (f \in E).$$

Each of them turns E into a normed space.

(d) Let $S \neq \varnothing$ be a set, and let (Y, d) be a metric space. A function $f \colon S \to Y$ is said to be *bounded* if

$$\sup_{x, y \in S} d(f(x), f(y)) < \infty$$

The set

$$B(S, Y) := \{f \colon S \to Y : f \text{ is bounded}\}$$

becomes a metric space through D defined by

$$D(f, g) := \sup_{x \in S} d(f(x), g(x)) \qquad (f, g \in B(S, Y)).$$

(e) France is a centralized country: every train that goes from one French city to another has to pass through Paris. This is slightly exaggerated, but not too much, as the map shows.

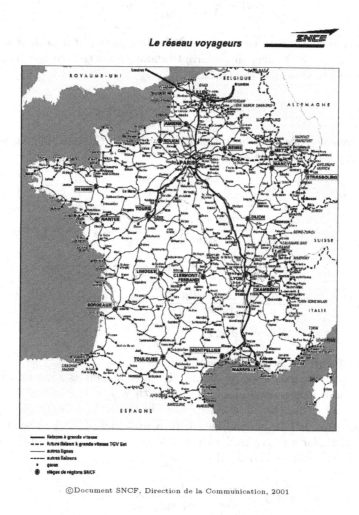

Fig. 2.1: Map of the French railroad network

This motivates the name *French railroad metric* for the following construction. Let (X, d) be a metric space ("France"), and fix $p \in X$ ("Paris"). Define a new metric d_p on X by letting

$$d_p(x, y) := \begin{cases} 0, & x = y, \\ d(x, p) + d(p, y), & \text{otherwise,} \end{cases}$$

for $x, y \in X$. Then (X, d_p) is again a metric space.

(f) Let (X, d) be any metric space, and define $\tilde{d} \colon X \times X \to \mathbb{R}$ via

$$\tilde{d}(x, y) := \frac{d(x, y)}{1 + d(x, y)} \qquad (x, y \in X).$$

We claim that $\tilde{d}$ is a metric on X. It is obvious that $\tilde{d}$ is positive definite and symmetric. Hence, all we have to show is that the triangle inequality holds. First note that the function

$$[0, \infty) \to \mathbb{R}, \quad t \mapsto \frac{t}{1 + t} \qquad\qquad (*)$$

is increasing (this can be verified, for instance, through differentiation). Let $x, y, z \in X$, and observe that

$$
\begin{aligned}
\tilde{d}(x, z) &= \frac{d(x, z)}{1 + d(x, z)} \\
&\leq \frac{d(x, y) + d(y, z)}{1 + d(x, y) + d(y, z)}, \qquad \text{because } (*) \text{ is increasing,} \\
&= \frac{d(x, y)}{1 + d(x, y) + d(y, z)} + \frac{d(y, z)}{1 + d(x, y) + d(y, z)} \\
&\leq \frac{d(x, y)}{1 + d(x, y)} + \frac{d(y, z)}{1 + d(y, z)} \\
&= \tilde{d}(x, y) + \tilde{d}(y, z).
\end{aligned}
$$

Consequently, $\tilde{d}$ is indeed a metric on X.

(g) A *semimetric* d on a set X satisfies the same axioms as a metric with one exception: it is possible for $x, y \in X$ with $x \neq y$ that $d(x, y) = 0$. If d is a semimetric, then $\tilde{d}$ as constructed in the previous example is also a semimetric. Let X be equipped with a sequence $(d_n)_{n=1}^{\infty}$ of semimetrics such that, for any $x, y \in X$ with $x \neq y$, there is $n \in \mathbb{N}$ with $d_n(x, y) > 0$. Then $d \colon X \times X \to \mathbb{R}$ defined by

$$d(x, y) := \sum_{n=1}^{\infty} \frac{1}{2^n} \frac{d_n(x, y)}{1 + d_n(x, y)} \qquad (x, y \in X)$$

is a metric. Clearly, d is symmetric and satisfies the triangle inequality, and if $x, y \in X$ are such that $x \neq y$, there is $n \in \mathbb{N}$ with $d_n(x, y) > 0$, so that $d(x, y) \geq \frac{1}{2^n} \frac{d_n(x,y)}{1 + d_n(x,y)} > 0$.

(h) The previous example can be used, for instance, to turn a Cartesian product X of countably many metric spaces $((X_n, d_n))_{n=1}^{\infty}$ into a metric space again. For each $n \in \mathbb{N}$, the map

$$\delta_n \colon X \times X \to [0, \infty), \quad ((x_1, x_2, x_3, \ldots), (y_1, y_2, y_3, \ldots)) \mapsto d_n(x_n, y_n)$$

is a semimetric. Moreover, if $x = (x_1, x_2, x_3, \ldots)$ and $y = (y_1, y_2, y_3, \ldots)$ are different points of X, there is at least one coordinate $n \in \mathbb{N}$ such that $x_n \neq y_n$, so that $\delta_n(x, y) = d_n(x_n, y_n) > 0$. For $x = (x_1, x_2, x_3, \ldots)$ and $y = (y_1, y_2, y_3, \ldots)$ in X, let

$$d(x, y) := \sum_{n=1}^{\infty} \frac{1}{2^n} \frac{\delta_n(x, y)}{1 + \delta_n(x, y)} = \sum_{n=1}^{\infty} \frac{1}{2^n} \frac{d_n(x_n, y_n)}{1 + d_n(x_n, y_n)}.$$

Then d is a metric on X.

(i) Let X be any set. For $x, y \in X$ define

$$d(x, y) := \begin{cases} 0, & x = y, \\ 1, & \text{otherwise.} \end{cases}$$

Then (X, d) is easily seen to be a metric space. (Metric spaces of this form are called *discrete*.)

Exercises

1. Let S be any set, and let X consist of the finite subsets of S. Show that

$$d \colon X \times X \to [0, \infty), \quad (A, B) \mapsto |(A \setminus B) \cup (B \setminus A)|$$

is a metric on X.

2. Verify Example 2.1.2(d) in detail.

3. Let $S \neq \varnothing$ be a set, and let E be a normed space. Show that

$$\|f\|_\infty := \sup\{\|f(x)\| : x \in S\} \quad (f \in B(S, E))$$

defines a norm on $B(S, E)$. How does $\|\cdot\|_\infty$ relate to the metric D from the previous exercise?

4. Let $(E, \|\cdot\|)$ be a normed space, and define $\|\|\cdot\|\| \colon E \to [0, \infty)$ by letting

$$\|\|x\|\| := \frac{\|x\|}{1 + \|x\|} \quad (x \in E).$$

Is $\|\|\cdot\|\|$ a norm on E?

5. Let X be any set, and let $d \colon X \times X \to [0, \infty)$ be a semimetric. For $x, y \in X$, define $x \approx y$ if and only if $d(x, y) = 0$.
   (a) Show that $\approx$ is an equivalence relation on X.
   (b) For $x \in X$, let $[x]$ denote its equivalence class with respect to $\approx$, and let $X/\approx$ denote the collection of all $[x]$ with $x \in X$. Show that

$$(X/\approx) \times (X/\approx) \to [0, \infty), \quad ([x], [y]) \mapsto d(x, y)$$

   defines a metric on $X/\approx$.

2.2 Open and Closed Sets

We start with the definition of an open ball in a metric space:

Definition 2.2.1. Let (X, d) be a metric space, let $x_0 \in X$, and let $r > 0$. The open ball *centered at* x_0 *with radius* r is defined as

$$B_r(x_0) := \{x \in X : d(x, x_0) < r\}.$$

Of course, in Euclidean 2- or 3-space, this definition coincides with the usual intuitive one. Nevertheless, even though open balls are defined with the intuitive notions of Euclidean space in mind, matters can turn out to be surprisingly counterintuitive:

Examples 2.2.2. (a) Let (X, d) be a discrete metric space, let $x_0 \in X$, and let $r > 0$. Then

$$B_r(x_0) = \begin{cases} \{x_0\}, & r < 1, \\ X, & r \geq 1, \end{cases}$$

holds; that is, each open ball is a singleton subset or the whole space.

(b) Let (X, d) be any metric space, let $p \in X$, and let d_p be the corresponding French railroad metric. To tell open balls in (X, d) and (X, d_p) apart, we write $B_r(x_0; d)$ and $B_r(x_0; d_p)$, respectively, for $x_0 \in X$ and $r > 0$. Let $x_0 \in X$, and let $r > 0$. Since, for $x \in X$ with $x \neq x_0$, we have

$$d_p(x, x_0) = d(x, p) + d(p, x_0) < r \quad \Longleftrightarrow \quad d(x, p) < r - d(p, x_0),$$

the following dichotomy holds.

$$B_r(x_0; d_p) = \begin{cases} \{x_0\}, & \text{if } r \leq d(p, x_0), \\ B_{r-d(p,x_0)}(p; d) \cup \{x_0\}, & \text{otherwise.} \end{cases}$$

Like the notion of an open ball, the notion of an open set extends from Euclidean space to arbitrary metric spaces.

Definition 2.2.3. Let (X, d) be a metric space. A set $U \subset X$ is called open if, for each $x \in U$, there is $\epsilon > 0$ such that $B_\epsilon(x) \subset U$.

If our choice of terminology is to make any sense, an open ball in a metric space better be an open set. Indeed, this is true.

Example 2.2.4. Let (X, d) be a metric space, let $x_0 \in X$, and let $r > 0$. For $x \in B_r(x_0)$, choose $\epsilon := r - d(x, x_0) > 0$. Hence, we have for $y \in B_\epsilon(x)$:

$$d(y, x_0) \leq d(y, x) + d(x, x_0) < \epsilon + d(x, x_0) = r - d(x, x_0) + d(x, x_0) = r.$$

It follows that $B_\epsilon(x) \subset B_r(x_0)$.

The following proposition lists the fundamental properties of open sets.

Proposition 2.2.5. Let (X, d) be a metric space. Then:

(i) $\varnothing$ and X are open;

(ii) If $\mathcal{U}$ is a family of open subsets of X, then $\bigcup\{U : U \in \mathcal{U}\}$ is open;

(iii) If U_1 and U_2 are open subsets of X, then $U_1 \cap U_2$ is open.

Proof. (i) is clear.

For (ii), let $\mathcal{U}$ be a family of open sets in X, and let $x \in \bigcup\{U : U \in \mathcal{U}\}$. Then there is $U_0 \in \mathcal{U}$ with $x \in U_0$, and since U_0 is open there is $\epsilon > 0$ such that

$$B_\epsilon(x) \subset U_0 \subset \bigcup\{U : U \in \mathcal{U}\}.$$

Hence, $\bigcup\{U : U \in \mathcal{U}\}$ is open.

Let $U_1, U_2 \subset X$ be open, and let $x \in U_1 \cap U_2$. Since U_1 and U_2 are open, there are $\epsilon_1, \epsilon_2 > 0$ such that $B_{\epsilon_j}(x) \subset U_j$ for $j = 1, 2$. Let $\epsilon := \min\{\epsilon_1, \epsilon_2\}$. Then it is immediate that $B_\epsilon(x) \subset U_1 \cap U_2$. This proves (iii). $\square$

Proposition 2.2.5(i) may seem odd at the first glance. The closed unit interval in $\mathbb{R}$ is a subspace of $\mathbb{R}$, thus a metric space in its own right, and thus open by Proposition 2.2.5(i). But, of course, we know that $[0, 1]$ is *not* open. How is this possible? The answer is that openness (as well as all the notions that are derived from it) depends on the context of a given metric space. Thus, $[0, 1]$ is open in $[0, 1]$, but not open in $\mathbb{R}$.

Example 2.2.6. Let (X, d) be a discrete metric space, and let $S \subset X$. Then

$$S = \bigcup_{x \in S} \{x\} = \bigcup_{x \in S} B_1(x)$$

is open; that is, all subsets of X are open.

A notion closely related to open sets is that of a neighborhood of a point.

Definition 2.2.7. *Let (X, d) be a metric space, and let $x \in X$. A subset N of X is called a* neighborhood *of x if there is an open subset U of X with $x \in U \subset N$. The collection of all neighborhoods of x is denoted by $\mathcal{N}_x$.*

Proposition 2.2.8. *Let (X, d) be a metric space, and let $x \in X$. Then:*

(i) *A subset N of X belongs to $\mathcal{N}_x$ if and only if there is $\epsilon > 0$ such that $B_\epsilon(x) \subset N$;*

(ii) *If $N \in \mathcal{N}_x$ and $M \supset N$, then $M \in \mathcal{N}_x$;*

(iii) *If $N_1, N_2 \in \mathcal{N}_x$, then $N_1 \cap N_2 \in \mathcal{N}_x$.*

Moreover, a subset U of X is open if and only if $U \in \mathcal{N}_y$ for each $y \in U$.

Proof. Suppose that $N \subset X$ is such that there is $\epsilon > 0$ such that $B_\epsilon(x) \subset N$. Since $B_\epsilon(x)$ is open, it follows that $N \in \mathcal{N}_x$. Conversely, suppose that $N \in \mathcal{N}_x$. Then there is an open subset U of N with $x \in U$. By the definition of openness, there is $\epsilon > 0$ such that $B_\epsilon(x) \subset U \subset N$. This proves (i).

(ii) is obvious, and (iii) follows immediately from Proposition 2.2.5(iii).

Let $U \subset X$ be open. Then, clearly, U is a neighborhood of each of its points. Conversely, let $U \subset X$ be any set with that property. By the definition of a neighborhood, there is, for each $y \in U$, an open subset U_y of U with $y \in U_y$. Since $U = \bigcup_{y \in U} U_y$, Proposition 2.2.5(ii) yields that U is open. $\square$

As in Euclidean space, we define a set to be closed if its complement is open.

Definition 2.2.9. *Let (X, d) be a metric space. A subset F of X is called closed if $X \setminus F$ is open.*

Examples 2.2.10. (a) Let (X, d) be any metric space, let $x_0 \in X$, and let $r > 0$. The *closed ball* centered at x_0 with radius r is defined as

$$B_r[x_0] := \{x \in X : d(x, x_0) \le r\}.$$

We claim that $B_r[x_0]$ is indeed closed. To show this, let $x \in X \setminus B_r[x_0]$, that is, such that $d(x, x_0) > r$. Let $\epsilon := d(x, x_0) - r > 0$, and let $y \in B_\epsilon(x)$. Since $d(x, x_0) \le d(x, y) + d(y, x_0)$, we obtain that

$$d(y, x_0) \ge d(x, x_0) - d(x, y) > d(x, x_0) - \epsilon = d(x, x_0) - (d(x, x_0) - r) = r.$$

It follows that $B_\epsilon(x) \subset X \setminus B_r[x_0]$. Consequently, $X \setminus B_r[x_0]$ is open and $B_r[x_0]$ is closed.

(b) In a discrete metric space, every subset is both open and closed.

The following is a straightforward consequence of Proposition 2.2.5.

Proposition 2.2.11. *Let (X, d) be a metric space. Then:*

(i) *$\varnothing$ and X are closed;*
(ii) *If $\mathcal{F}$ is a family of closed subsets of X, then $\bigcap \{F : F \in \mathcal{F}\}$ is closed;*
(iii) *If F_1 and F_2 are closed subsets of X, then $F_1 \cup F_2$ is closed.*

Of course, in most metric spaces there are many sets that are neither open nor closed. Nevertheless, we can make the following definition.

Definition 2.2.12. *Let (X, d) be a metric space. For each $S \subset X$, the closure of S is defined as*

$$\overline{S} := \bigcap \{F : F \subset X \text{ is closed and contains } S\}.$$

From Proposition 2.2.11(ii) it is immediate that the closure of a set is a closed set. The following is an alternative description of the closure.

Proposition 2.2.13. *Let (X, d) be a metric space, and let $S \subset X$. Then we have:*

$$\overline{S} = \{x \in X : N \cap S \ne \varnothing \text{ for all } N \in \mathcal{N}_x\}$$
$$= \{x \in X : B_\epsilon(x) \cap S \ne \varnothing \text{ for all } \epsilon > 0\}.$$

Proof. Each open ball is a neighborhood of its center, and any neighborhood of a point contains an open ball centered at that point; therefore

$$\{x \in X : N \cap S \neq \varnothing \text{ for all } N \in \mathcal{N}_x\} = \{x \in X : B_\epsilon(x) \cap S \neq \varnothing \text{ for all } \epsilon > 0\}$$

holds. We denote this set by cl(S).

Let $x \in \overline{S}$, and let $N \in \mathcal{N}_x$. Then there is an open subset U of X contained in N with $x \in U$. Assume that $N \cap S = \varnothing$, so that $U \cap S = \varnothing$ (i.e., $S \subset X \setminus U$). Since $X \setminus U$ is closed, it follows that $\overline{S} \subset X \setminus U$ and thus $x \in X \setminus U$, which is a contradiction. Consequently, $x \in$ cl(S) holds.

Conversely, let $x \in$ cl(S), and assume that $x \notin \overline{S}$. Then $U := X \setminus \overline{S}$ is an open set containing x (thus belonging to $\mathcal{N}_x$) having empty intersection with S. This contradicts $x \in$ cl(S). $\square$

Examples 2.2.14. (a) Any open interval in $\mathbb{R}$ contains a rational number. Hence, we have $\overline{\mathbb{Q}} = \mathbb{R}$.

(b) Let (X, d) be any metric space. It is obvious that $\overline{B_r(x_0)} \subset B_r[x_0]$ for all $x_0 \in X$ and $r > 0$. In general, equality need not hold. If (X, d) is discrete and has more than one element, we have for any $x_0 \in X$ that

$$\overline{B_1(x_0)} = \overline{\{x_0\}} = \{x_0\} \subsetneq X = B_1[x_0].$$

(c) Let E be a normed space, let $x_0 \in E$, and let $r > 0$. We claim that (in this particular situation) $\overline{B_r(x_0)} = B_r[x_0]$ holds. In view of the previous example, only $B_r[x_0] \subset \overline{B_r(x_0)}$ needs proof. Let $x \in B_r[x_0]$, and let $\epsilon > 0$. Choose $\delta \in (0, 1)$ such that $\delta \|x - x_0\| < \epsilon$, and let

$$y := x_0 + (1 - \delta)(x - x_0) = (1 - \delta)x + \delta x_0,$$

so that

$$\|y - x_0\| = (1 - \delta)\|x - x_0\| \leq (1 - \delta)r < r;$$

that is, $y \in B_r(x_0)$. Furthermore, we have

$$\|y - x\| = \|(1 - \delta)x + \delta x_0 - x\| = \delta \|x - x_0\| < \epsilon,$$

and thus $y \in B_\epsilon(x)$. From Proposition 2.2.13, we conclude that $x \in \overline{B_r(x_0)}$.

The closure of a set is important in connection with two further topological concepts: density and the boundary.

Definition 2.2.15. *Let (X, d) be a metric space.*

(a) *A subset D of X is said to be* dense *in X if $\overline{D} = X$.*
(b) *If X has a dense countable subset, then X is called* separable.

Examples 2.2.16. (a) $\mathbb{Q}$ is dense in $\mathbb{R}$. In particular, $\mathbb{R}$ is separable.

(b) A subset S of a discrete metric space (X, d) is dense if and only if $S = X$. In particular, X is separable if and only if it is countable.

The following hereditary property of separability is somewhat surprising, but very useful.

Theorem 2.2.17. *Let (X, d) be a separable metric space, and let Y be a subspace of X. Then Y is also separable.*

Proof. Let $C = \{x_1, x_2, x_3, \dots\}$ be a dense countable subset of X. One might be tempted to use $Y \cap C$ as a dense (and certainly countable) subset of Y, but this may not work: if $X \neq C$, take $Y = X \setminus C$, for example.

Let

$$\mathbb{A} := \left\{ (n, m) \in \mathbb{N} \times \mathbb{N} : \text{there is } y \in Y \text{ such that } d(y, x_n) < \frac{1}{m} \right\}.$$

For each $(n, m) \in \mathbb{A}$, choose $y_{n,m} \in Y$ with $d(y_{n,m}, x_n) < \frac{1}{m}$. Then $C_Y := \{y_{n,m} : (n, m) \in \mathbb{A}\}$ is a countable subset of Y. We claim that C_Y is also dense in Y. Let $y \in Y$, and let $\epsilon > 0$. Choose $m \in \mathbb{N}$ such that $\frac{1}{m} \leq \frac{\epsilon}{2}$. Since C is dense in X, there is $n \in \mathbb{N}$ such that $d(y, x_n) < \frac{1}{m}$. By the definition of $\mathbb{A}$, this means that $(n, m) \in \mathbb{A}$. It follows that

$$d(y, y_{n,m}) \leq d(y, x_n) + d(x_n, y_{n,m}) < \frac{2}{m} \leq \epsilon.$$

By Proposition 2.2.13, this means that y lies in the closure of C_Y in Y. □

Examples 2.2.18. (a) The irrational numbers are a separable subspace of $\mathbb{R}$.
(b) Let $X = B(\mathbb{N}, \mathbb{R})$ be equipped with the metric introduced in Example 2.1.2(d); that is,

$$d(f, g) = \sup_{n \in \mathbb{N}} |f(n) - g(n)| \qquad (f, g \in X).$$

We claim that X is not separable. We assume towards a contradiction that X *is* separable. Let Y denote the subspace of X consisting of all $\{0, 1\}$ valued functions. From Theorem 2.2.17, it follows that Y is separable, too. Since, for $f, g \in Y$, we have

$$d(f, g) = \sup_{n \in \mathbb{N}} |f(n) - g(n)| = \begin{cases} 0, & f = g, \\ 1, & f \neq g, \end{cases}$$

it follows that Y is a discrete metric space and therefore must be countable. However, the map

$$Y \to [0, 1], \quad f \mapsto \sum_{n=1}^{\infty} \frac{f(n)}{2^n}$$

is surjective, and $[0, 1]$ is not countable. This is a contradiction.

To motivate the notion of boundary, we first consider an example.

Example 2.2.19. Let $(E, \| \cdot \|)$ be a normed space, let $x_0 \in E$, and let $r > 0$. Then, intuitively, one might view the boundary of the open ball $B_r(x_0)$ as the *sphere*

$$S_r[x_0] := \{x \in E : \|x - x_0\| = r\}.$$

Let $x \in S_r[x_0]$, and let $\epsilon > 0$. Let $\delta \in (0,1)$ be such that $\delta \|x - x_0\| < \epsilon$, and let $y := x_0 + (1 - \delta)(x - x_0)$. As in Example 2.2.14(c), it follows that $y \in B_\epsilon(x) \cap B_r(x_0)$, so that

$$B_\epsilon(x) \cap B_r(x_0) \neq \varnothing \qquad \text{and} \qquad B_\epsilon(x) \cap (E \setminus B_r(x_0)) \neq \varnothing. \qquad (**)$$

On the other hand, since $B_r(x_0)$ and $E \setminus B_r[x_0]$ are open, it follows that any element x of E satisfying $(**)$ for each $\epsilon > 0$ must lie in $S_r[x_0]$.

In view of this example, we define the following.

Definition 2.2.20. *Let (X, d) be a metric space, and let $S \subset X$. Then the boundary of S is defined as*

$$\partial S := \{x \in X : B_\epsilon(x) \cap S \neq \varnothing \text{ and } B_\epsilon(x) \cap (X \setminus S) \neq \varnothing \text{ for all } \epsilon > 0\}.$$

An argument similar to that at the beginning of the proof of Proposition 2.2.13 yields immediately that

$$\partial S = \{x \in X : N \cap S \neq \varnothing \text{ and } N \cap (X \setminus S) \neq \varnothing \text{ for all } N \in \mathcal{N}_x\}$$

for each subset S of a metric space X.

Proposition 2.2.21. *Let (X, d) be a metric space, and let $S \subset X$. Then:*

(i) $\partial S = \partial(X \setminus S)$;
(ii) ∂S *is closed;*
(iii) $\overline{S} = S \cup \partial S$.

Proof. (i) is a triviality.

For (ii), let $x \in X \setminus \partial S$; that is, there is $N \in \mathcal{N}_x$ such that $N \cap S = \varnothing$ or $N \cap (X \setminus S) = \varnothing$. Let $U \subset N$ be open such that $x \in U$. It follows that $U \cap S = \varnothing$ or $U \cap (X \setminus S) = \varnothing$. Since U is a neighborhood of each of its points, it follows that $U \subset X \setminus \partial S$. Hence, $X \setminus \partial S$ is a neighborhood of x. Since x was arbitrary, it follows that $X \setminus \partial S$ is open.

For (iii), note that, by Proposition 2.2.13, $\partial S \subset \overline{S}$ holds, so that $S \cup \partial S \subset \overline{S}$. Conversely, let $x \in \overline{S}$, and suppose that $x \notin S$. For each $N \in \mathcal{N}_x$, it is clear that $N \cap (X \setminus S) \neq \varnothing$, and Proposition 2.2.13 yields that $N \cap S \neq \varnothing$ as well. $\square$

The closure of a given set is, by definition, the smallest closed set containing it. Analogously, one defines the largest open set contained in a given set.

Definition 2.2.22. *Let (X,d) be a metric space. For each $S \subset X$, the* interior *of S is defined as*

$$\overset{\circ}{S} := \bigcup \{U : U \subset X \text{ is open and contained in } S\}.$$

The following proposition characterizes the interior of a set:

Proposition 2.2.23. *Let (X,d) be a metric space, and let $S \subset X$. Then we have:*

$$\overset{\circ}{S} = \{x \in X : S \in \mathcal{N}_x\} = S \setminus \partial S.$$

Proof. Let $x \in \overset{\circ}{S}$. Then there is an open subset U of S with $x \in U$, so that $S \in \mathcal{N}_x$. Conversely, if $S \in \mathcal{N}_x$, then there is an open set U of X with $x \in U \subset S$, so that $x \in \overset{\circ}{S}$.

Let $x \in \overset{\circ}{S}$, so that $S \in \mathcal{N}_x$ by the foregoing. Since, trivially, $S \cap (X \setminus S) = \varnothing$, we see that $x \notin \partial S$. Conversely, let $x \in S \setminus \partial S$. Then there is $N \in \mathcal{N}_x$ such that $N \cap (X \setminus S) = \varnothing$. Let $U \subset N$ be open in X such that $x \in U$. It follows that $U \cap (X \setminus S) = \varnothing$ and therefore $U \subset S$. Consequently, $x \in U \subset \overset{\circ}{S}$ holds.
□

Exercises

1. Show that a finite subset of a metric space is closed.
2. Let $(E, \|\cdot\|)$ be a normed space, let $U \subset E$ be open, and let $S \subset E$ be any set. Show that $S + U := \{x + y : x \in S, y \in U\}$ is open in E.
3. Let $U \subset \mathbb{R}$ be open.
   (a) For each $x \in U$, let I_x be the union of all open intervals contained in U and containing x. Show that I_x is an open (possibly unbounded) interval.
   (b) For $x, y \in U$, show that $I_x = I_y$ or $I_x \cap I_y = \varnothing$.
   (c) Conclude that U is a union of countably many, pairwise disjoint open intervals.
4. Let (X,d) be a metric space, and let $S \subset X$. The *distance* of $x \in X$ to S is defined as
$$\text{dist}(x, S) := \inf\{d(x,y) : y \in S\}$$
   (where $\text{dist}(x, S) = \infty$ if $S = \varnothing$). Show that $\overline{S} = \{x \in X : \text{dist}(x, S) = 0\}$.
5. Let Y be the subspace of $B(\mathbb{N}, \mathbb{F})$ consisting of those sequences tending to zero. Show that Y is separable.
6. Let (X,d) be a metric space, and let Y be a subspace of X. Show that $U \subset Y$ is open in Y if and only if there is $V \subset X$ that is open in X such that $U = Y \cap V$.

2.3 Convergence and Continuity

The notion of convergence in $\mathbb{R}^n$ carries over to metric spaces almost verbatim.

Definition 2.3.1. *Let (X,d) be a metric space. A sequence $(x_n)_{n=1}^\infty$ in X is said to* converge *to $x \in X$ if, for each $\epsilon > 0$, there is $n_\epsilon \in \mathbb{N}$ such that $d(x_n, x) < \epsilon$ for all $n \geq n_\epsilon$. We then say that x is the* limit *of $(x_n)_{n=1}^\infty$ and write $x = \lim_{n\to\infty} x_n$ or $x_n \to x$.*

It is straightforward to verify that a sequence $(x_n)_{n=1}^\infty$ in a metric space converges to x if and only if, for each $N \in \mathcal{N}_x$, there is $n_N \in \mathbb{N}$ such that $x_n \in N$ for all $n \geq n_N$.

Examples 2.3.2. (a) Let (X,d) be a discrete metric space, and let $(x_n)_{n=1}^\infty$ be a sequence in X that converges to $x \in X$. Then there is $n_1 \in \mathbb{N}$ such that $d(x_n, x) < 1$ for $n \geq n_1$; that is, $x_n = x$ for $n \geq n_1$. Hence, every convergent sequence in a discrete metric space is eventually constant.

(b) Let $C([0,1], \mathbb{F})$ be equipped with the metric induced by $\|\cdot\|_\infty$ (Example 2.1.2(c)). We claim that a sequence $(f_n)_{n=1}^\infty$ in $C([0,1], \mathbb{F})$ converges to $f \in C([0,1], \mathbb{F})$ with respect to that metric if and only if it converges (to f) uniformly on $[0,1]$. Suppose first that $\|f_n - f\|_\infty \to 0$, and let $\epsilon > 0$. Then there is $n_\epsilon \in \mathbb{N}$ such that

$$|f_n(t) - f(t)| \leq \|f_n - f\|_\infty < \epsilon \qquad (n \geq n_\epsilon, t \in [0,1]),$$

so that $f_n \to f$ uniformly on $[0,1]$. Conversely, let $(f_n)_{n=1}^\infty$ converge to f uniformly on $[0,1]$, and let $\epsilon > 0$. By the definition of uniform convergence, there is $n_\epsilon \in \mathbb{N}$ such that

$$|f_n(t) - f(t)| < \frac{\epsilon}{2} \qquad (n \geq n_\epsilon, t \in [0,1])$$

and consequently,

$$\|f_n - f\|_\infty = \sup\{|f_n(t) - f(t)| : t \in [0,1]\} \leq \frac{\epsilon}{2} < \epsilon \qquad (n \geq n_\epsilon).$$

Hence, we have convergence with respect to $\|\cdot\|_\infty$.

As in $\mathbb{R}^n$, the limit of a sequence in a metric space is unique.

Proposition 2.3.3. *Let (X,d) be a metric space, let $(x_n)_{m=1}^\infty$ be a sequence in X, and let $x, x' \in X$ be such that $(x_n)_{n=1}^\infty$ converges to both x and x'. Then x and x' are equal.*

Proof. Assume that $x \neq x'$, so that $\epsilon := \frac{1}{2}d(x,x') > 0$. Since $x_n \to x$, there is $n_1 \in \mathbb{N}$ such that $d(x_n, x) < \epsilon$ for $n \geq n_1$, and since $x_n \to x'$, too, there is $n_2 \in \mathbb{N}$ such that $d(x_n, x') < \epsilon$ for $n \geq n_2$. Let $n := \max\{n_1, n_2\}$, so that

$$d(x, x') \leq d(x, x_n) + d(x_n, x') < \epsilon + \epsilon = d(x, x'),$$

which is nonsense. $\square$

Here is the idea of the proof of Proposition 2.3.3 in a sketch.

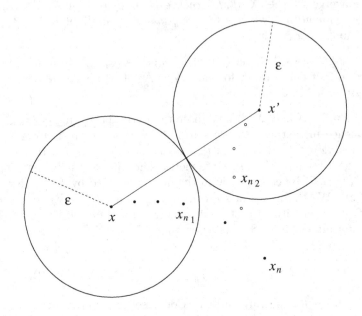

Fig. 2.2: Uniqueness of the limit

Also, as in $\mathbb{R}^n$, convergence in metric spaces can be used to characterize the closed subsets.

Proposition 2.3.4. *Let (X, d) be a metric space, and let $S \subset X$. Then $\overline{S}$ consists of those points in X that are the limit of a sequence in S.*

Proof. Let $x \in X$ be the limit of a sequence $(x_n)_{n=1}^{\infty}$ in S, and let $\epsilon > 0$. By the definition of convergence, there is $n_{\epsilon} \in \mathbb{N}$ such that $d(x_n, x) < \epsilon$ for $n \geq n_{\epsilon}$; that is, $x_n \in B_{\epsilon}(x)$ for $n \geq n_{\epsilon}$. In particular, $B_{\epsilon}(x) \cap S$ is nonempty. Since $\epsilon > 0$ is arbitrary, it follows that $x \in \overline{S}$ by Proposition 2.2.13.

Conversely, let $x \in \overline{S}$. By Proposition 2.2.13, we have $B_{\frac{1}{n}}(x) \cap S \neq \varnothing$ for each $n \in \mathbb{N}$; there is thus, for each $n \in \mathbb{N}$, some $x_n \in S$ with $d(x_n, x) < \frac{1}{n}$. It is clear that the sequence $(x_n)_{n=1}^{\infty}$ converges to x. $\square$

Corollary 2.3.5. *Let (X, d) be a metric space. Then $F \subset X$ is closed if and only if every sequence in F that converges in X has its limit in F.*

Of course, with a notion of convergence at hand, continuity of functions can be defined.

Definition 2.3.6. *Let (X, d_X) and (Y, d_Y) be metric spaces, and let $x_0 \in X$. Then $f \colon X \to Y$ is said to be continuous at x_0 if, for each sequence $(x_n)_{n=1}^{\infty}$ in X that converges to x_0, we have $\lim_{n \to \infty} f(x_n) = f(x_0)$.*

The following characterization holds.

Theorem 2.3.7. *Let (X, d_X) and (Y, d_Y) be metric spaces, and let $x_0 \in X$. Then the following are equivalent for $f \colon X \to Y$.*

(i) *f is continuous at x_0.*
(ii) *For each $\epsilon > 0$, there is $\delta > 0$ such that $d_Y(f(x), f(x_0)) < \epsilon$ for all $x \in X$ with $d_X(x, x_0) < \delta$.*
(iii) *For each $\epsilon > 0$, there is $\delta > 0$ such that $B_\delta(x_0) \subset f^{-1}(B_\epsilon(f(x_0)))$.*
(iv) *For each $N \in \mathcal{N}_{f(x_0)}$, we have $f^{-1}(N) \in \mathcal{N}_{x_0}$.*

Proof. (i) $\implies$ (ii): Assume otherwise; that is, there is $\epsilon_0 > 0$ such that, for each $\delta > 0$, there is $x_\delta \in X$ with $d_X(x_\delta, x_0) < \delta$, but $d_Y(f(x_\delta), f(x_0)) \geq \epsilon_0$. For $n \in \mathbb{N}$, let $x_n' := x_{\frac{1}{n}}$, so that $d(x_n', x_0) < \frac{1}{n}$ and thus $x_n' \to x_0$. Since, however, $d_Y(f(x_n'), f(x_0)) \geq \epsilon_0$ holds for all $n \in \mathbb{N}$, it is impossible that $f(x_n') \to f(x_0)$ as required for f to be continuous at x_0.

(iii) is only a rewording of (ii).

(iii) $\implies$ (iv): Let $N \in \mathcal{N}_{f(x_0)}$. Hence, there is $\epsilon > 0$ such that $B_\epsilon(x_0) \subset N$. By (iii), there is $\delta > 0$ such that

$$B_\delta(x_0) \subset f^{-1}(B_\epsilon(f(x_0))) \subset f^{-1}(N).$$

This implies that $f^{-1}(N) \in \mathcal{N}_{x_0}$.

(iv) $\implies$ (i): Let $(x_n)_{n=1}^\infty$ be a sequence in X with $x_n \to x_0$. Let $N \in \mathcal{N}_{f(x_0)}$, so that $f^{-1}(N) \in \mathcal{N}_{x_0}$. Since $x_n \to x_0$, there is $n_N \in \mathbb{N}$ such that $x_n \in f^{-1}(N)$ for $n \geq n_N$; that is, $f(x_n) \in N$ for $n \geq n_N$. Since $N \in \mathcal{N}_{f(x_0)}$ was arbitrary, this yields $f(x_n) \to f(x_0)$. $\square$

The following definition should also look familiar.

Definition 2.3.8. *Let (X, d_X) and (Y, d_Y) be metric spaces. Then a function $f \colon X \to Y$ is said to be* continuous *if it is continuous at each point of X.*

Example 2.3.9. Let (X, d) be a metric space. We first claim that

$$|d(x, y) - d(x_0, y_0)| \leq d(x, x_0) + d(y, y_0) \qquad (x, x_0, y, y_0 \in X). \qquad (\ast\ast\ast)$$

Fix $x, x_0, y, y_0 \in X$, and note that

$$d(x, y) \leq d(x, x_0) + d(x_0, y_0) + d(y_0, y)$$

and therefore

$$d(x, y) - d(x_0, y_0) \leq d(x, x_0) + d(y_0, y).$$

Interchanging the roles of x and x_0 and, respectively, y and y_0, yields $d(x_0, y_0) - d(x, y) \leq d(x, x_0) + d(y_0, y)$. Altogether, we obtain $(\ast\ast\ast)$. The Cartesian square X^2 becomes a metric space in its own right through

$$\tilde{d}((x, y), (x', y')) := d(x, x') + d(y, y') \qquad ((x, x'), (y, y') \in X^2).$$

The inequality $(\ast\ast\ast)$ immediately yields that $d \colon X^2 \to \mathbb{R}$ is continuous if X^2 is equipped with $\tilde{d}$.

Corollary 2.3.10. *Let (X, d_X) and (Y, d_Y) be metric spaces. Then the following are equivalent for $f \colon X \to Y$.*

(i) *f is continuous.*
(ii) *$f^{-1}(U)$ is open in X for each open subset U of Y.*
(iii) *$f^{-1}(F)$ is closed in X for each closed subset F of Y.*

Proof. (i) $\implies$ (ii): Let $U \subset Y$ be open, so that $U \in \mathcal{N}_y$ for each $y \in U$ and thus $U \in \mathcal{N}_{f(x)}$ for each $x \in f^{-1}(U)$. For Theorem 2.3.7(iv), we conclude that $f^{-1}(U) \in \mathcal{N}_x$ for each $x \in f^{-1}(U)$; that is, $f^{-1}(U)$ is a neighborhood of each of its points and thus open.

(ii) $\implies$ (iii): Let $F \subset Y$ be closed, so that $Y \setminus F$ is open. Since $X \setminus f^{-1}(F) = f^{-1}(Y \setminus F)$ then must be open by (ii), it follows that $f^{-1}(F)$ is closed. Analogously, (iii) $\implies$ (ii) is proved.

(ii) $\implies$ (i): If f satisfies (ii), it trivially also satisfies Theorem 2.3.7(iii) for each $x \in X$. $\square$

We now give an example which shows that continuous maps between general metric spaces can be quite different from what we may intuitively expect.

Example 2.3.11. Let (X, d_X) and (Y, d_Y) be metric spaces such that (X, d_X) is discrete, and let $f \colon X \to Y$ be arbitrary. Let $U \subset Y$ be open. Since in a discrete space every set is open, it follows that $f^{-1}(U)$ is open. Consequently, f must be continuous.

As we have seen, there can be different metrics on one set. For many purposes, it is convenient to view certain metrics as identical.

Definition 2.3.12. *Let X be a set. Two metrics d_1 and d_2 on X are said to be* equivalent *if the identity map on X is continuous both from (X, d_1) to (X, d_2) and from (X, d_2) to (X, d_1).*

In view of Corollary 2.3.10, two metrics d_1 and d_2 on a set X are equivalent if and only if they yield the same open sets (or, equivalently, the same closed sets).

Examples 2.3.13. (a) The Euclidean metric on $\mathbb{R}^n$ and the discrete metric are not equivalent.
(b) For $j = 1, \ldots, n$, let (X_j, d_j) be a metric space. Let $X := X_1 \times \cdots \times X_n$, and for $x = (x_1, \ldots, x_n), y = (y_1, \ldots, y_n) \in X$, define

$$D_1(x, y) := \sum_{j=1}^{n} d_j(x_j, y_j) \qquad \text{and} \qquad D_\infty(x, y) := \max_{j=1,\ldots,n} d_j(x_j, y_j).$$

Then D_1 and D_∞ are metrics on X satisfying

$$D_\infty(x, y) \le D_1(x, y) \le n \, D_\infty(x, y) \qquad (x, y \in X).$$

Consequently, D_1 and D_∞ are equivalent.

(c) Let (X, d) be any metric space, let $p \in X$, and let d_p be the corresponding French railroad metric. Since

$$d(x, y) \le d_p(x, y) \qquad (x, y \in X),$$

it is easily seen that the identity is continuous from (X, d_p) to (X, d). On the other hand, let $(x_n)_{n=1}^{\infty}$ be a sequence in X that converges to $x \ne p$ with respect to d. If $x_n \ne x$, we have

$$d_p(x_n, x) = d(x_n, p) + d(p, x) \ge d(p, x),$$

so that, for $d_p(x_n, x) \to 0$ to hold, $(x_n)_{n=1}^{\infty}$ must be eventually constant. Hence, for example, the Euclidean metric on $\mathbb{R}^n$ and—no matter how "Paris" is chosen—the corresponding French railroad metric are not equivalent. On the other hand, if (X, d) is discrete, then the identity from (X, d) to (X, d_p) is also continuous, so that d and d_p are equivalent.

(d) Let (X, d) be any metric space, and let $\tilde{d}$ be the metric defined in Example 2.1.2(f). We claim that d and $\tilde{d}$ are equivalent. The function

$$f \colon [0, \infty) \to [0, 1), \quad t \mapsto \frac{t}{1+t}$$

is continuous and bijective with continuous inverse

$$g \colon [0, 1) \to [0, \infty), \quad s \mapsto \frac{s}{1-s}.$$

Since $\tilde{d} = f \circ d$ (and, consequently, $d = g \circ \tilde{d}$), it follows that d and $\tilde{d}$ are indeed equivalent.

(e) Let (X, d) be any metric space, and let $U \subset X$ be open. Define

$$d_U(x, y) := d(x, y) + \left| \frac{1}{\operatorname{dist}(x, X \setminus U)} - \frac{1}{\operatorname{dist}(y, X \setminus U)} \right| \qquad (x, y \in U).$$

(If $U = X$, we formally set $\frac{1}{\operatorname{dist}(x, X \setminus U)} = \frac{1}{\operatorname{dist}(y, X \setminus U)} = \frac{1}{\infty} = 0$.) From Exercise 2.2.4, it follows that d_U is well defined on $U \times U$. We claim that d_U is a metric on U. Clearly, d_U is positive definite and symmetric. Let $x, y, z \in U$, and note that

$$\begin{aligned}
d_U(x, z) &= d(x, z) + \left| \frac{1}{\operatorname{dist}(x, X \setminus U)} - \frac{1}{\operatorname{dist}(z, X \setminus U)} \right| \\
&\le d(x, y) + d(y, z) + \left| \frac{1}{\operatorname{dist}(x, X \setminus U)} - \frac{1}{\operatorname{dist}(y, X \setminus U)} \right. \\
&\qquad\qquad\qquad \left. + \frac{1}{\operatorname{dist}(y, X \setminus U)} - \frac{1}{\operatorname{dist}(z, X \setminus U)} \right| \\
&\le d(x, y) + \left| \frac{1}{\operatorname{dist}(x, X \setminus U)} - \frac{1}{\operatorname{dist}(y, X \setminus U)} \right| \\
&\qquad + d(y, z) + \left| \frac{1}{\operatorname{dist}(y, X \setminus U)} - \frac{1}{\operatorname{dist}(z, X \setminus U)} \right| \\
&= d_U(x, y) + d_U(y, z).
\end{aligned}$$

We claim that d restricted to $U \times U$ and d_U are equivalent. Since

$$d(x, y) \leq d_U(x, y) \qquad (x, y \in U),$$

the continuity of the identity from (U, d_U) to (U, d) is clear. To prove that the identity on U is also continuous in the converse direction, first note that nothing has to be shown if $U = X$. We may thus suppose without loss of generality that $U \subsetneq X$. Let $(x_n)_{n=1}^{\infty}$ be a sequence in U that converges to $x \in U$ with respect to d; that is, $d(x_n, x) \to 0$. By Exercise 3 below, this entails that $\mathrm{dist}(x_n, X \setminus U) \to \mathrm{dist}(x, X \setminus U)$ and thus

$$d_U(x_n, x) = d(x_n, x) + \left| \frac{1}{\mathrm{dist}(x_n, X \setminus U)} - \frac{1}{\mathrm{dist}(x, X \setminus U)} \right| \to 0.$$

Hence, $(x_n)_{n=1}^{\infty}$ converges to x as well with respect to d_U.

Exercises

1. Let $((X_k, d_k))_{k=1}^{\infty}$ be a sequence of metric spaces, and let $X := \prod_{k=1}^{\infty} X_k$ be equipped with the metric d from Example 2.1.2(g). Show that convergence in X is coordinatewise convergence: a sequence $\left(\left(x_1^{(n)}, x_2^{(n)}, x_3^{(n)}, \ldots \right) \right)_{n=1}^{\infty}$ in X converges to $(x_1, x_2, x_3, \ldots) \in X$ with respect to d if and only if $x_k^{(n)} \to x_k$ for each $k \in \mathbb{N}$.
2. Let (X, d_X) and (Y, d_Y) be metric spaces, let $p \in X$, and let d_p denote the corresponding French railroad metric on X. Show that $f \colon X \to Y$ is continuous with respect to d_p if and only if it is continuous at p with respect to d_X.
3. Let (X, d) be a metric space, and let $\varnothing \neq S \subset X$. Show that the function

$$X \to \mathbb{R}, \qquad x \mapsto \mathrm{dist}(x, S)$$

   is continuous.
4. Let E and F be normed spaces, and let $T \colon E \to F$ be linear. Show that the following are equivalent.
   (i) T is continuous;
   (ii) T is continuous at 0;
   (iii) There is $C \geq 0$ such that $\|T(x)\| \leq C\|x\|$ for all $x \in E$.
5. Let E and F be normed spaces, let $T \colon E \to F$ be linear, and suppose that $\dim E < \infty$. Show that T is continuous. (*Hint:* For $x \in E$, define $\|\|x\|\| := \max\{\|x\|, \|T(x)\|\}$; show that $\|\| \cdot \|\|$ is a norm on E, and use Proposition B.1.)
6. On $C([0, 1], \mathbb{F})$ we have the two norms $\| \cdot \|_1$ and $\| \cdot \|_{\infty}$ introduced in Example 2.1.2(c). Show that the metrics induced by these two norms are not equivalent.

2.4 Completeness

As we can define convergent sequences in metric spaces, we can speak of Cauchy sequences.

Definition 2.4.1. *Let (X, d) be a metric space. A sequence $(x_n)_{n=1}^{\infty}$ in X is called a* Cauchy sequence *if, for each $\epsilon > 0$, there is $n_\epsilon > 0$ such that $d(x_n, x_m) < \epsilon$ for all $n, m \geq n_\epsilon$.*

As in $\mathbb{R}^n$, we have the following.

Proposition 2.4.2. *Let (X, d) be a metric space, and let $(x_n)_{n=1}^{\infty}$ be a convergent sequence in X. Then $(x_n)_{n=1}^{\infty}$ is a Cauchy sequence.*

Proof. Let $x := \lim_{n \to \infty} x_n$, and let $\epsilon > 0$. Then there is $n_\epsilon > 0$ such that $d(x_n, x) < \frac{\epsilon}{2}$ for all $n \geq n_\epsilon$. Consequently, we have

$$d(x_n, x_m) \leq d(x_n, x) + d(x, x_m) < \frac{\epsilon}{2} + \frac{\epsilon}{2} = \epsilon \qquad (n, m \geq n_\epsilon),$$

so that $(x_n)_{n=1}^{\infty}$ is a Cauchy sequence. $\square$

In $\mathbb{R}^n$, the converse holds as well: Every Cauchy sequence converges. For general metric spaces, this is clearly false: the sequence $\left(\frac{1}{n}\right)_{n=1}^{\infty}$ is a Cauchy sequence in the metric space $(0, 1)$—equipped with its canonical metric—but has no limit in that space. This makes the following definition significant.

Definition 2.4.3. *A metric space (X, d) is called* complete *if every Cauchy sequence in X converges.*

A normed space that is complete with respect to the metric induced by its norm is also called a *Banach space*.

Examples 2.4.4. (a) $\mathbb{R}^n$ is complete.
(b) In a discrete metric space, every Cauchy sequence is eventually constant and therefore convergent. Hence, discrete metric spaces are complete.
(c) Let $S \neq \varnothing$ be a set, and let (Y, d) be a complete metric space. We claim that the metric space $(B(S, Y), D)$ from Example 2.1.2(d) is complete. Let $(f_n)_{n=1}^{\infty}$ be a Cauchy sequence in $B(S, Y)$. Let $\epsilon > 0$, and choose $n_\epsilon > 0$ such that $D(f_n, f_m) < \epsilon$ for all $n, m \geq n_\epsilon$. For $x \in S$, we then have

$$d(f_n(x), f_m(x)) \leq D(f_n, f_m) < \epsilon \qquad (n, m \geq n_\epsilon).$$

Consequently, $(f_n(x))_{n=1}^{\infty}$ is a Cauchy sequence in Y for each $x \in S$. Since Y is complete, we can therefore define $f \colon S \to Y$ by letting

$$f(x) := \lim_{n \to \infty} f_n(x) \qquad (x \in S).$$

We first claim that f lies in $B(S, Y)$ and is, in fact, the limit of $(f_n)_{n=1}^{\infty}$ with respect to D. To see this, let $x \in S$, and note that

$$d(f_n(x), f(x)) = \lim_{m \to \infty} d(f_n(x), f_m(x))$$

for $n \in \mathbb{N}$ by Example 2.3.9. It follows for $n \geq n_\epsilon$ that

$$d(f_n(x), f(x))$$

$$= \lim_{m \to \infty} d(f_n(x), f_m(x)) \le \limsup_{m \to \infty} D(f_n, f_m) \le \epsilon \qquad (x \in S).$$

Let $n \ge n_\epsilon$, and let $C := \sup_{x,y \in S} d(f_n(x), f_n(y))$, which is finite by the definition of $B(S, Y)$. From the previous inequality, we obtain, for arbitrary $x, y \in S$, that

$$d(f(x), f(y)) \le d(f(x), f_n(x)) + d(f_n(x), f_n(y)) + d(f_n(y), f(y)) \le 2\epsilon + C.$$

Hence, f belongs to $B(S, Y)$. Since $d(f_n(x), f(x)) \le \epsilon$ for all $x \in S$ and $n \ge n_\epsilon$, we eventually obtain:

$$D(f_n, f) = \sup_{x \in S} d(f_n(x), f(x)) \le \epsilon \qquad (n \ge n_\epsilon).$$

This is sufficient to guarantee that $f = \lim_{n \to \infty} f_n$ in $(B(S, Y), D)$.

The following proposition indicates how to get new complete spaces from old ones.

Proposition 2.4.5. *Let (X, d) be a metric space, and let Y be a subspace of X.*

(i) *If X is complete and if Y is closed in X, then Y is complete.*
(ii) *If Y is complete, then it is closed in X.*

Proof. Suppose that X is complete and that Y is closed in X. Let $(x_n)_{n=1}^{\infty}$ be a Cauchy sequence in Y. Then $(x_n)_{n=1}^{\infty}$ is also a Cauchy sequence in X and thus has a limit $x \in X$. Since Y is closed, Corollary 2.3.5 yields that $x \in Y$, so that Y is complete. This proves (i).

For (ii), let $(y_n)_{n=1}^{\infty}$ be a sequence in Y that converges to $y \in X$. Since $(y_n)_{n=1}^{\infty}$ converges in X, it is a Cauchy sequence in X and thus in Y. Since Y is complete, there is $y' \in Y$ with $y' = \lim_{n \to \infty} y_n$. If $(y_n)_{n=1}^{\infty}$ converges to y' in Y, it does so in X. Uniqueness of the limit yields that $y' = y$. Hence, y lies in Y. Corollary 2.3.5 thus yields that Y is closed in X. $\square$

Example 2.4.6. Let (X, d_X) and (Y, d_Y) be metric spaces. We define

$$C(X, Y) := \{f : X \to Y : f \text{ is continuous}\}$$

and

$$C_b(X, Y) := B(X, Y) \cap C(X, Y).$$

Clearly, $C_b(X, Y)$ is a subspace of the metric space $(B(X, Y), D)$. We claim that $C_b(X, Y)$ is closed in $B(X, Y)$ and therefore complete if (Y, d_Y) is. Let $(f_n)_{n=1}^{\infty}$ be a sequence in $C_b(X, Y)$ that converges to $f \in B(X, Y)$. We claim that f is again continuous. To see this, fix $x_0 \in X$. We show that f is continuous at x_0. Let $\epsilon > 0$. Since $f_n \to f$ in $B(X, Y)$, there is $n_\epsilon \in \mathbb{N}$ such that

$D(f_n, f) < \frac{\epsilon}{3}$ for $n \geq n_\epsilon$. Fix $n \geq n_\epsilon$. Since f_n is continuous at x_0, the set $N := f_n^{-1}(B_{\frac{\epsilon}{3}}(f_n(x_0)))$ is a neighborhood of x_0. Let $x \in N$, and note that

$$
\begin{aligned}
d(f(x), f(x_0)) &\leq d(f(x), f_n(x)) + d(f_n(x), f_n(x_0)) + d(f_n(x_0), f(x_0)) \\
&\leq D(f_n, f) + d(f_n(x), f_n(x_0)) + D(f_n, f) \\
&< \frac{2\epsilon}{3} + d(f_n(x), f_n(x_0)), \qquad \text{because } n \geq n_\epsilon, \\
&< \epsilon, \qquad \text{because } x \in N.
\end{aligned}
$$

It follows that $N \subset f^{-1}(B_\epsilon(f(x_0)))$, so that $f^{-1}(B_\epsilon(f(x_0))) \in \mathcal{N}_{x_0}$. Since $\epsilon > 0$ was arbitrary, this is enough to guarantee the continuity of f at x_0.

In view of Proposition 2.4.5, the following assertion seems to defy reason at first glance.

Proposition 2.4.7. *Let (X, d) be a complete metric space, and let $U \subset X$ be open. Then (U, d_U) is a complete metric space, where d_U is defined as in Example 2.3.13(e).*

Proof. If $U = X$, we have $d_U = d$, so that the claim is trivially true. Hence, suppose that $U \subsetneq X$.

Let $(x_n)_{n=1}^\infty$ be a Cauchy sequence in (U, d_U). Then $(x_n)_{n=1}^\infty$ is easily seen to be a Cauchy sequence in (X, d) as well. Let $x \in X$ be its limit in (X, d). We first claim that $x \in U$. Assume towards a contradiction that $x \in X \setminus U$. From Exercise 2.3.3, we conclude that $\operatorname{dist}(x_n, X \setminus U) \to 0$. Since $(x_n)_{n=1}^\infty$ is a Cauchy sequence in (U, d_U), there is $n_1 \in \mathbb{N}$ such that

$$
\left| \frac{1}{\operatorname{dist}(x_n, X \setminus U)} - \frac{1}{\operatorname{dist}(x_m, X \setminus U)} \right| \leq d_U(x_n, x_m) \leq 1 \qquad (n, m \geq n_1).
$$

Fix $m \geq n_1$, and note that therefore

$$
\begin{aligned}
\frac{1}{\operatorname{dist}(x_n, X \setminus U)} &\leq \left| \frac{1}{\operatorname{dist}(x_n, X \setminus U)} - \frac{1}{\operatorname{dist}(x_m, X \setminus U)} \right| + \frac{1}{\operatorname{dist}(x_m, X \setminus U)} \\
&\leq 1 + \frac{1}{\operatorname{dist}(x_m, X \setminus U)} \qquad (n \geq n_1).
\end{aligned}
$$

This is impossible, however, if $\operatorname{dist}(x_n, X \setminus U) \to 0$. Consequently, $x \in U$ must hold.

Since d and d_U are equivalent on U, we see that $d_U(x_n, x) \to 0$ as well. Hence, x is the limit of $(x_n)_{n=1}^\infty$ in (U, d_U). $\square$

At first glance, Proposition 2.4.7 seems to be paradoxical, to say the least. Any open subset of a complete metric space is supposed to be complete with respect to an equivalent metric. Doesn't this and Proposition 2.4.5(ii) immediately yield that every open subset of a complete metric space is also closed?

This is clearly wrong. The apparent paradox is resolved if one recalls the definition of a sub*space* of a metric space: (U, d_U) is not a subspace of the metric space (X, d), even though the two metrics d and d_U are equivalent on U.

We now present a famous property of complete metric spaces, for which we first require a definition.

Definition 2.4.8. *Let* (X, d) *be a metric space. The* diameter *of a subset* $S \neq \varnothing$ *of* X *is defined as*

$$\mathrm{diam}(S) := \sup\{d(x, y) : x, y \in S\}.$$

Theorem 2.4.9 (Cantor's intersection theorem). *Let* (X, d) *be a complete metric space, and let* $(F_n)_{n=1}^{\infty}$ *be a sequence of nonempty closed subsets of* X *such that* $F_1 \supset F_2 \supset F_3 \supset \cdots$ *and* $\lim_{n \to \infty} \mathrm{diam}(F_n) = 0$. *Then* $\bigcap_{n=1}^{\infty} F_n$ *contains precisely one point of* X.

Proof. For each $n \in \mathbb{N}$, let $x_n \in F_n$. We claim that the sequence $(x_n)_{n=1}^{\infty}$ is a Cauchy sequence. To see this, let $\epsilon > 0$. Choose $n_\epsilon \in \mathbb{N}$ such that $\mathrm{diam}(F_n) < \epsilon$ for $n \geq n_\epsilon$. Let $n, m \geq n_\epsilon$. Since the sequence $(F_n)_{n=1}^{\infty}$ is decreasing, it follows that $x_n, x_m \in F_{n_\epsilon}$, so that

$$d(x_n, x_m) \leq \mathrm{diam}(F_{n_\epsilon}) < \epsilon.$$

Consequently, $(x_n)_{n=1}^{\infty}$ is indeed a Cauchy sequence and therefore converges in X, to x say. Since $x_m \in F_m \subset F_n$ for all $n, m \in \mathbb{N}$, with $m \geq n$, it follows from Corollary 2.3.5 that $x = \lim_{m \to \infty} x_m \in F_n$ for all $n \in \mathbb{N}$ and thus $x \in \bigcap_{n=1}^{\infty} F_n$.

To show that $\bigcap_{n=1}^{\infty} F_n = \{x\}$, assume towards a contradiction that there is $x' \in \bigcap_{n=1}^{\infty} F_n$ different from x. Let $\epsilon_0 := d(x, x') > 0$, and choose $n \in \mathbb{N}$ so large that $\mathrm{diam}(F_n) < \epsilon_0$. Since $x, x' \in F_n$, we obtain

$$d(x, x') \leq \mathrm{diam}(F_n) < \epsilon_0 = d(x, x'),$$

which is impossible. $\square$

Next, we show that *any* metric space is—in a sense yet to be made precise—already a subspace of a complete metric space.

Definition 2.4.10. *Let* (X, d) *be a metric space. A* completion *of* (X, d) *is a metric space* $\left(\tilde{X}, \tilde{d}\right)$ *together with a map* $\iota : X \to \tilde{X}$ *with the following properties.*

(a) $\left(\tilde{X}, \tilde{d}\right)$ *is complete;*

(b) $\tilde{d}(\iota(x), \iota(y)) = d(x, y)$ *for* $x, y \in X$;

(c) $\iota(X)$ *is dense in* $\tilde{X}$.

We show that, first of all, *every* metric space has a completion and, secondly, that this completion is unique (in a certain sense).

To specify what we mean by uniqueness of a completion, we require another definition.

Definition 2.4.11. *Let (X, d_X) and (X, d_Y) be metric spaces. A function f: $X \to Y$ is called an* isometry *(or* isometric*) if*

$$d_Y(f(x), f(y)) = d_X(x, y) \qquad (x, y \in X).$$

If f is also bijective, we call f an isometric isomorphism.

Lemma 2.4.12. *Let (X, d) be a metric space, let $\left(\tilde{X}_1, \tilde{d}_1\right)$ and $\left(\tilde{X}_2, \tilde{d}_2\right)$ be completions of (X, d), and let $\iota_1 : X \to \tilde{X}_1$ and $\iota_2 : X \to \tilde{X}_2$ denote the corresponding maps from Definition 2.4.10. Then there is a unique isometric isomorphism $f : \tilde{X}_1 \to \tilde{X}_2$ such that $f \circ \iota_1 = \iota_2$.*

Proof. We begin with the definition of f. Let $x \in \tilde{X}_1$. Since $\iota_1(X)$ is dense in $\tilde{X}_1$, there is a sequence $(x_n)_{n=1}^\infty$ in X such that $x = \lim_{n \to \infty} \iota_1(x_n)$. It is clear that $(\iota_1(x_n))_{n=1}^\infty$ is a Cauchy sequence in $\tilde{X}_1$, and Definition 2.4.10(b) implies that $(x_n)_{n=1}^\infty$ is a Cauchy sequence in X. Again Definition 2.4.10(b) guarantees that $(\iota_2(x_n))_{n=1}^\infty$ is a Cauchy sequence in $\tilde{X}_2$ and therefore converges. Let $f(x) := \lim_{n \to \infty} \iota_2(x_n)$.

We first prove that f is *well defined*, that is, does not depend on the particular choice of a sequence $(x_n)_{n=1}^\infty$. To prove this, let $(x'_n)_{n=1}^\infty$ be another sequence in X with $x = \lim_{n \to \infty} \iota_1(x'_n)$. It follows that

$$d(x_n, x'_n) = \tilde{d}_1(\iota_1(x_n), \iota_1(x'_n)) \le \tilde{d}_1(\iota_1(x_n), x) + \tilde{d}_1(x, \iota_1(x'_n)) \to 0$$

and therefore

$$
\begin{aligned}
\tilde{d}_2(\iota_2(x'_n), f(x)) &\le \tilde{d}_2(\iota_2(x'_n), \iota_2(x_n)) + \tilde{d}_2(\iota_2(x_n), f(x)) \\
&= d(x'_n, x_n) + \tilde{d}_2(\iota_2(x_n), f(x)) \\
&\to 0.
\end{aligned}
$$

All in all, $f(x) = \lim_{n \to \infty} \iota_2(x'_n)$ holds, so that f is indeed well defined.

Next, we prove that f is an isometry. Let $x, y \in \tilde{X}_1$ and let $(x_n)_{n=1}^\infty$ and $(y_n)_{n=1}^\infty$ be the corresponding sequences in X used to define $f(x)$ and $f(y)$, respectively. From

$$
\begin{aligned}
\tilde{d}_2(f(x), f(y)) &= \lim_{n \to \infty} \tilde{d}_2(\iota_2(x_n), \iota_2(x_n)) \\
&= \lim_{n \to \infty} d(x_n, y_n) \\
&= \lim_{n \to \infty} \tilde{d}_1(\iota_1(x_n), \iota_1(x_n)) \\
&= \tilde{d}_1(x, y),
\end{aligned}
$$

we see that f is isometric. This immediately also proves the injectivity of f.

Clearly, $f \circ \iota_1 = \iota_2$ holds, so that $f\left(\tilde{X}_1\right) \supset \iota_2(X)$ must be dense in $\tilde{X}_2$. We claim that $f\left(\tilde{X}_1\right)$ is a complete subspace of $\tilde{X}_2$ and therefore closed (this implies that $f\left(\tilde{X}_1\right)$ must be all of $\tilde{X}_2$). Let $(x_n)_{n=1}^\infty$ be a sequence in $\tilde{X}_1$ such that $(f(x_n))_{n=1}^\infty$ is a Cauchy sequence in $\tilde{X}_2$. Since f is an isometry, $(x_n)_{n=1}^\infty$ is also a Cauchy sequence in $\tilde{X}_1$ and thus convergent to some $x \in \tilde{X}_1$. Again since f is an isometry, it follows that $\lim_{n\to\infty} f(x_n) = f(x)$ in $\tilde{X}_2$.

Finally, to prove the uniqueness of f, let $\tilde{f}\colon \tilde{X}_1 \to \tilde{X}_2$ be another map as described in the statement of the lemma. Let $x \in \tilde{X}_1$. By Definition 2.4.10(c), there is a sequence $(x_n)_{n=1}^\infty$ in X with $\lim_{n\to\infty} \iota_1(x_n) = x$. We obtain that

$$f(x) = \lim_{n\to\infty} f(\iota_1(x_n)) = \lim_{n\to\infty} \iota_2(x_n) = \lim_{n\to\infty} \tilde{f}(\iota_1(x_n)) = \tilde{f}(x).$$

Since $x \in \tilde{X}_1$ was arbitrary, this proves that $f = \tilde{f}$. $\quad\square$

In less formal (but probably much more digestible) language, Lemma 2.4.12 asserts that a completion of a metric space (if it exists at all!) is unique up to isometric isomorphism.

The existence of the completion of a given metric space is surprisingly easy to establish.

Theorem 2.4.13. *Let (X, d) be a metric space. Then (X, d) has a completion, which is unique up to isometric isomorphism.*

Proof. In view of Lemma 2.4.12, only the existence of the completion still has to be shown. It is sufficient to find some complete metric space and an isometry ι from X into that space: just let $\tilde{X} := \overline{\iota(X)}$. The complete metric space into which we embed X is the Banach space $C_b(X, \mathbb{R})$.

Fix $x_0 \in X$. For $x \in X$, define

$$f_x\colon X \to \mathbb{R}, \quad t \mapsto d(x, t) - d(x_0, t).$$

In view of Example 2.3.9, it is clear that f_x is continuous for each $x \in X$, and also, due to the inequality $(***)$ from Example 2.3.9, we have

$$|f_x(t)| \leq d(x, x_0) + d(t, t) = d(x, x_0) \quad (t \in X),$$

so that f_x lies even in $C_b(X, \mathbb{R})$. We claim that the map

$$\iota\colon X \to C_b(X, \mathbb{R}), \quad x \mapsto f_x$$

is an isometry. To see this, fix $x, y \in X$ and note that, by $(***)$ again,

$$D(\iota(x), \iota(y)) = \sup_{t \in X} |f_x(t) - f_y(t)| = \sup_{t \in X} |d(x, t) - d(y, t)| \leq d(x, y),$$

holds; on the other hand, we have

$$D(\iota(x), \iota(y)) = \sup_{t \in X} |f_x(t) - f_y(t)| \geq |f_x(y) - f_y(y)| = d(x, y),$$

which proves the claim. $\quad\square$

In view of the uniqueness of a completion up to isometric isomorphism, we are justified to speak of *the* completion of a metric space. For the sake of notational convenience, we also identify a metric space with its canonical image in its completion.

We now turn to one of the most fundamental theorems on complete metric spaces.

Theorem 2.4.14 (Bourbaki's Mittag-Leffler theorem). *Suppose that* $((X_n, d_n))_{n=0}^{\infty}$ *is a sequence of complete metric spaces, and let* $f_n \colon X_n \to X_{n-1}$ *for* $n \in \mathbb{N}$ *be continuous with dense range. Then*

$$\bigcap_{n=1}^{\infty} (f_1 \circ f_2 \circ \cdots \circ f_n)(X_n)$$

is dense in X_0.

Proof. We first inductively define new metrics $\tilde{d}_0, \tilde{d}_1, \tilde{d}_2, \ldots$ on the spaces $X_0, X_1, X_2, \ldots$ such that

- $\tilde{d}_n$ and d_n are equivalent for $n \in \mathbb{N}_0$,
- $\left(X_n, \tilde{d}_n \right)$ is complete for each $n \in \mathbb{N}_0$, and
- $\tilde{d}_{n-1}(f_n(x), f_n(y)) \leq \tilde{d}_n(x, y)$ for $n \in \mathbb{N}$ and $x, y \in X_n$.

This is accomplished by letting $\tilde{d}_0 := d_0$ and, once $\tilde{d}_0, \ldots, \tilde{d}_{n-1}$ have been defined for some $n \in \mathbb{N}$, letting

$$\tilde{d}_n(x, y) := d_n(x, y) + \tilde{d}_{n-1}(f_n(x), f_n(y)) \qquad (x, y \in X_n).$$

In what follows, we consider the spaces $X_0, X_1, X_2, \ldots$ equipped with the metrics $\tilde{d}_0, \tilde{d}_1, \tilde{d}_2, \ldots$ instead of with $d_0, d_1, d_2, \ldots$.

Let $U_0 \subset X$ be open and not empty. We need to show that

$$U_0 \cap \bigcap_{n=1}^{\infty} (f_1 \circ \cdots \circ f_n)(X_n) \neq \varnothing.$$

Since $f_1(X_1)$ is dense in X_0, there is $x_1 \in X_1$ with $\overline{f_1(x_1)} \in U_0$. Since f_1 is continuous at x_1, there is $\delta_1 \in (0, 1]$ such that $\overline{f_1(B_{\delta_1}(x_1))} \subset U_0$. Let $U_1 := B_{\delta_1}(x_1)$. Since $f_2(X_2)$ is dense in X_1, there is $x_2 \in X_2$ with $\overline{f_2(x_2)} \in U_1$. Since f_2 is continuous at x_2, there is $\delta_2 \in \left(0, \frac{1}{2} \right]$ such that $\overline{f_2(B_{\delta_2}(x_2))} \subset U_1$. Let $U_2 := B_{\delta_2}(x_2)$, and continue in this fashion.

We thus obtain a sequence $(U_n)_{n=1}^{\infty}$ of open balls such that $\overline{f_n(U_n)} \subset U_{n-1}$ for $n \in \mathbb{N}$ and such that U_n has radius at most $\frac{1}{n}$. For $n \in \mathbb{N}_0$ and $m \in \mathbb{N}$, let

$$Y_{n,m} := \overline{(f_{n+1} \circ \cdots \circ f_{n+m})(U_{n+m})}.$$

It follows that $Y_{n,m} \neq \varnothing$, that $\operatorname{diam}(Y_{n,m}) \leq \frac{2}{n+m}$, and that $Y_{n,m+1} \subset Y_{n,m}$. From Cantor's intersection theorem, it follows that there is $y_n \in \bigcap_{m=1}^{\infty} Y_{n,m}$.

From the construction, it is immediate that $f_n(y_n) = y_{n-1}$ and thus that $(f_1 \circ \cdots \circ f_n)(y_n) = y_0$ for $n \in \mathbb{N}$. Consequently,

$$y_0 \in U_0 \cap \bigcap_{n=1}^{\infty} (f_1 \circ \cdots \circ f_n)(X_n)$$

holds.   □

The name "Mittag-Leffler theorem" for Theorem 2.4.14 may sound bewildering, but the well-known Mittag-Leffler theorem from complex analysis (Theorem A.1) can be obtained as a consequence of it (see Appendix A; besides some background from complex variables, you will also need material from Sections 3.1 to 3.4 for it). We turn, however, to another consequence of Theorem 2.4.14.

Lemma 2.4.15. *Let (X, d) be a metric space, and let $U_1, \ldots, U_n \subset X$ be dense open subsets of X. Then $U_1 \cap \cdots \cap U_n$ is dense in X.*

Proof. By induction, it is clear that we may limit ourselves to the case where $n = 2$. Let $x \in X$, and let $\epsilon > 0$. Since U_1 is dense in X, we have $B_\epsilon(x) \cap U_1 \neq \varnothing$. Since $B_\epsilon(x) \cap U_1$ is open—and thus a neighborhood of each of its points—it follows from the denseness of U_2 that $B_\epsilon(x) \cap U_1 \cap U_2 \neq \varnothing$. Since $\epsilon > 0$ was arbitrary, we conclude that $x \in \overline{U_1 \cap U_2}$.   □

Theorem 2.4.16 (Baire's theorem).   *Let (X, d) be a complete metric space, and let $(U_n)_{n=1}^{\infty}$ be a sequence of dense open subsets of X. Then $\bigcap_{n=1}^{\infty} U_n$ is dense in X.*

Proof. By Lemma 2.4.15, we may replace U_n by $U_1 \cap \cdots \cap U_n$ and thus suppose without loss of generality that $U_1 \supset U_2 \supset \cdots$. Let $(X_0, d_0) := (X, d)$, and let $(X_n, d_n) := (U_n, d_{U_n})$, where d_{U_n} is defined for $n \in \mathbb{N}$ as in Example 2.3.13(e). Furthermore, let $f_n \colon X_n \to X_{n-1}$ be the inclusion map for $n \in \mathbb{N}$. Since d and d_{U_n} are equivalent on X_n for $n \in \mathbb{N}$, it is clear that $f_1, f_2, \ldots$ are continuous. By the hypothesis, (X_0, d_0) is complete and the same is true for (X_n, d_n) with $n \in \mathbb{N}$ by Proposition 2.4.7. It follows from Theorem 2.4.14 that

$$\bigcap_{n=1}^{\infty} (f_1 \circ \cdots \circ f_n)(X_n) = \bigcap_{n=1}^{\infty} U_n$$

is dense in $X_0 = X$.   □

The following is an immediate consequence of Baire's theorem (just pass to complements).

Corollary 2.4.17. *Let (X, d) be a complete metric space, and let $(F_n)_{n=1}^{\infty}$ be a sequence of closed subsets of X such that $\bigcup_{n=1}^{\infty} F_n$ has a nonempty interior. Then at least one of the sets $F_1, F_2, \ldots$ has a nonempty interior.*

To illustrate the power of Baire's theorem, we turn to an example from elementary calculus. We all know that there are continuous functions that are not differentiable at certain points (take the absolute value function, for instance), and it is not very hard to come up with continuous functions that are not differentiable at a finite, and even countable, number of points. But is there a continuous function, on an interval say, that fails to be differentiable at each point of its domain? The following example gives the answer.

Example 2.4.18. For $n \in \mathbb{N}$, let F_n consist of those $f \in C([0,2], \mathbb{R})$ for which there is $t \in [0,1]$ such that

$$\sup_{h \in (0,1)} \frac{|f(t+h) - f(t)|}{h} \le n.$$

Obviously, if $f \in C([0,2], \mathbb{R})$ is differentiable at some point $t \in [0,1]$, then

$$\sup_{h \in (0,1)} \frac{|f(t+h) - f(t)|}{h} < \infty$$

must hold, so that $f \in \bigcup_{n=1}^{\infty} F_n$. Hence, if every continuous function on $[0,2]$ is differentiable at some point of $[0,1]$, we have $C([0,2], \mathbb{R}) = \bigcup_{n=1}^{\infty} F_n$. Using Corollary 2.4.17, we show that this is not possible.

To be able to apply Corollary 2.4.17, we first need to show that the sets F_n for $n \in \mathbb{N}$ are closed in $C([0,2], \mathbb{R})$. Fix $n \in \mathbb{N}$, and let $(f_m)_{m=1}^{\infty}$ be a sequence in F_n such that $\|f_m - f\|_\infty \to 0$ for some $f \in C([0,2], \mathbb{R})$. For each $m \in \mathbb{N}$, there is $t_m \in [0,1]$ such that

$$\sup_{h \in (0,1)} \frac{|f_m(t_m + h) - f_m(t_m)|}{h} \le n.$$

Suppose without loss of generality that $(t_m)_{m=1}^{\infty}$ converges to some $t \in [0,1]$ (otherwise, replace $(t_m)_{m=1}^{\infty}$ by a convergent subsequence). Fix $h \in (0,1)$ and $\epsilon > 0$, and choose $m_\epsilon \in \mathbb{N}$ so large that

$$\left\{ \begin{array}{c} |f(t+h) - f(t_m + h)| \\ \|f - f_m\|_\infty \\ |f(t_m) - f(t)| \end{array} \right\} < \frac{\epsilon}{4} h \qquad (m \ge m_\epsilon).$$

For $m \ge m_\epsilon$, this implies

$$|f(t+h) - f(t)|$$
$$\le \underbrace{|f(t+h) - f(t_m + h)|}_{< \frac{\epsilon}{4} h} + \underbrace{|f(t_m + h) - f_m(t_m + h)|}_{< \frac{\epsilon}{4} h}$$
$$+ \underbrace{|f_m(t_m + h) - f_m(t_m)|}_{\le nh} + \underbrace{|f_m(t_m) - f(t_m)|}_{< \frac{\epsilon}{4} h} + \underbrace{|f(t_m) - f(t)|}_{< \frac{\epsilon}{4} h}$$
$$\le nh + \epsilon h,$$

so that

$$\frac{|f(t+h) - f(t)|}{h} \leq n + \epsilon.$$

Since h and ϵ were arbitrary, this means that $f \in F_n$. Hence, F_n is closed.

Assume towards a contradiction that every $f \in C([0,2], \mathbb{R})$ is differentiable at some point in $[0,1]$, so that $C([0,2], \mathbb{R}) = \bigcup_{n=1}^{\infty} F_n$. By Corollary 2.4.17, there are $n_0 \in \mathbb{N}$, $f \in C([0,2], \mathbb{R})$, and $\epsilon > 0$ such that $B_\epsilon(f) \subset F_{n_0}$. By the Weierstraß approximation theorem (Corollary 4.3.8 below), $B_\epsilon(f)$ contains at least one polynomial, say p. Since $B_\epsilon(f)$ is open, there is $\delta > 0$ such that $B_\delta(p) \subset B_\epsilon(f) \subset F_{n_0}$. Replacing f by p and ϵ by δ, we can thus suppose without loss of generality that f is continuously differentiable on $[0,2]$.

For $k \in \mathbb{N}$ and $j = 0, \ldots, k$, let $t_j := \frac{2j}{k}$. Define a "sawtooth function" $g_k \colon [0,2] \to \mathbb{R}$ by letting

$$g_k(t) := \begin{cases} \frac{\epsilon}{2}k(t - t_{j-1}), \ t \in \left[t_{j-1}, t_{j-1} + \frac{1}{k}\right], \\ \frac{\epsilon}{2}k(t_j - t), \ \ t \in \left[t_j - \frac{1}{k}, t_j\right] \end{cases}$$

for $j = 1, \ldots, n$ and $t \in [t_{j-1}, t_j]$.

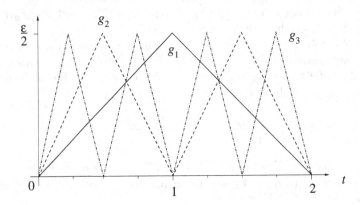

Fig. 2.3: Sawtooth functions

Then g_k is continuous with $\|g_k\|_\infty = \frac{\epsilon}{2}$, but

$$\sup_{h \in (0,1)} \frac{|g_k(t+h) - g_k(t)|}{h} = \frac{\epsilon}{2}k \tag{†}$$

holds for any $t \in [0,1]$. Since $f + g_k \in B_\epsilon(f) \subset F_{n_0}$, there is $t \in [0,1]$ such that

$$\sup_{h \in (0,1)} \frac{|(f+g_k)(t+h) - (f+g_k)(t)|}{h} \leq n_0.$$

This, however, yields

$$\sup_{h\in(0,1)} \frac{|g_k(t+h) - g_k(t)|}{h}$$

$$\leq \sup_{h\in(0,1)} \frac{|(f+g_k)(t+h) - (f+g_k)(t)|}{h} + \sup_{h\in(0,1)} \frac{|f(t+h) - f(t)|}{h}$$

$$= n_0 + \|f'\|_\infty,$$

which contradicts (†) if we choose $k \in \mathbb{N}$ so large that $\frac{\epsilon}{2}k > n_0 + \|f'\|_\infty$.

Hence, the sets $F_1, F_2, \ldots$ have an empty interior, thus their union $\bigcup_{n=1}^\infty F_n$ cannot be all of $C([0,2],\mathbb{R})$, and consequently there must be a continuous function on $[0,1]$ that is nowhere differentiable.

Exercises

1. Let (X,d) be any metric space, let $p \in X$, and let d_p be the corresponding French railroad metric. Show that (X, d_p) is complete.
2. Let (X,d) be a complete metric space, and let $(x_n)_{n=0}^\infty$ be a sequence in X such that there is $\theta \in (0,1)$ with $d(x_{n+1}, x_n) \leq \theta\, d(x_n, x_{n-1})$ for $n \in \mathbb{N}$. Show that $(x_n)_{n=0}^\infty$ is convergent.
3. Use the previous problem to prove *Banach's fixed point theorem*: if (X,d) is a complete metric space, and if $f\colon X \to X$ is such that

$$d(f(x), f(y)) \leq \theta\, d(x,y) \qquad (x, y \in X)$$

   for some $\theta \in (0,1)$, then there is a unique $x \in X$ with $f(x) = x$.
4. Let (X,d) be a metric space, and let $\varnothing \neq S \subset X$. Show that

$$\mathrm{diam}(S) = \inf\{r > 0 : S \subset B_r(x) \text{ for all } x \in S\}.$$

5. Give an example showing that the demand that $\lim_{n\to\infty} \mathrm{diam}(F_n) = 0$ in Cantor's intersection theorem cannot be dropped if we still want $\bigcap_{n=1}^\infty F_n \neq \varnothing$ to hold.
6. Let E be a normed space with a countable Hamel basis. Show that E is a Banach space if and only if $\dim E < \infty$. (*Hint:* You may use the fact that all finite-dimensional subspaces of a normed space are closed (Corollary B.3); then use Corollary 2.4.17.)
7. Let $(f_k)_{k=1}^\infty$ be a sequence in $C([0,1], \mathbb{F})$ that converges *pointwise* to a function $f\colon [0,1] \to \mathbb{F}$.
   (a) For $\theta > 0$ and $n \in \mathbb{N}$, let

$$F_n := \{t \in [0,1] : |f_n(t) - f_k(t)| \leq \theta \text{ for all } k \geq n\}.$$

   Show that F_n is closed, and that $[0,1] = \bigcup_{n=1}^\infty F_n$.
   (b) Let $\epsilon > 0$, and let I be a nondegenerate, closed subinterval of $[0,1]$. Show that there is a nondegenerate, closed interval J contained in $\overset{\circ}{I}$ such that

$$|f(t) - f(s)| \leq \epsilon \qquad (t, s \in J).$$

   (*Hint:* Apply (a) with $\theta := \frac{\epsilon}{3}$ and Corollary 2.4.17.)

(c) Let I be a nondegenerate closed subinterval of $[0, 1]$. Show that there is a sequence $(I_n)_{n=1}^{\infty}$ of nondegenerate closed subintervals of I with $I_1 \supset \overset{\circ}{I}_1 \supset I_2 \supset \overset{\circ}{I}_2 \supset I_3 \supset \cdots$ such that
  - The length of I_n is at most $\frac{1}{n}$, and
  - $|f(t) - f(s)| \leq \frac{1}{n}$ for all $s, t \in I_n$.

What can be said about f at all points in $\bigcap_{n=1}^{\infty} I_n$?

(d) Conclude that the set of points in $[0, 1]$ at which f is continuous is dense in $[0, 1]$.

2.5 Compactness for Metric Spaces

The notion of compactness is one of the most crucial in all of topology (and one of the hardest to grasp).

Definition 2.5.1. *Let (X, d) be a metric space, and let $S \subset X$. An* open cover *for S is a collection $\mathcal{U}$ of open subsets of X such that $S \subset \bigcup \{U : U \in \mathcal{U}\}$.*

Definition 2.5.2. *A subset K of a metric space (X, d) is called* compact *if, for each open cover $\mathcal{U}$ of K, there are $U_1, \ldots, U_n \in \mathcal{U}$ such that $K \subset U_1 \cup \cdots \cup U_n$.*

Definition 2.5.2 is often worded as, "A set is compact if and only if each open cover has a finite subcover."

Examples 2.5.3. (a) Let (X, d) be a metric space, and let $S \subset X$ be finite; that is, $S = \{x_1, \ldots, x_n\}$. Let $\mathcal{U}$ be an open cover of X. Then, for each $j = 1, \ldots, n$, there is $U_j \in \mathcal{U}$ such that $x_j \in U_j$. It follows that $S \subset U_1 \cup \cdots \cup U_n$. Hence, S is compact.

(b) Let (X, d) be a compact metric space, and let $\varnothing \neq K \subset X$ be compact. Fix $x_0 \in K$. Since $\{B_r(x_0) : r > 0\}$ is an open cover of K, there are $r_1, \ldots, r_n > 0$ such that

$$K \subset B_{r_1}(x_0) \cup \cdots \cup B_{r_n}(x_0).$$

With $R := \max\{r_1, \ldots, r_n\}$, we see that $K \subset B_R(x_0)$, so that $\text{diam}(K) \leq 2R < \infty$. This means, for example, that any unbounded subset of $\mathbb{R}^n$ (or, more generally, of any normed space) cannot be compact. In particular, the only compact normed space is $\{0\}$.

(c) Let $X = (0, 1)$ be equipped with the usual metric. For $r \in (0, 1)$, let $U_r := (r, 1)$. Then $\{U_r : r \in (0, 1)\}$ is an open cover for $(0, 1)$ which has no finite subcover.

Before we turn to more (and more interesting) examples of compact metric spaces, we establish a few hereditary properties.

Proposition 2.5.4. *Let (X, d) be a metric space, and let Y be a subspace of X.*

(i) *If X is compact and Y is closed in X, then Y is compact.*
(ii) *If Y is compact, then it is closed in X.*

Proof. For (i), let $\mathcal{U}$ be an open cover for Y. Since Y is closed in X, the family $\mathcal{U} \cup \{X \setminus Y\}$ is an open cover for X. Since X is compact, it has a finite subcover, i.e., there are $U_1, \ldots, U_n \in \mathcal{U}$ such that

$$X = U_1 \cup \cdots \cup U_n \cup X \setminus Y.$$

Taking the intersection with Y, we see that $Y \subset U_1 \cup \cdots \cup U_n$.

For (ii), let $x \in X \setminus Y$. For each $y \in Y$, there are $\epsilon_y, \delta_y > 0$ such that $B_{\epsilon_y}(x) \cap B_{\delta_y}(y) = \varnothing$. Since $\{B_{\delta_y}(y) : y \in Y\}$ is an open cover for Y, there are $y_1, \ldots, y_1 \in Y$ such that

$$Y \subset B_{\delta_{y_1}}(y_1) \cup \cdots \cup B_{\delta_{y_n}}(y_n).$$

Letting $\epsilon := \min\{\epsilon_{y_1}, \ldots, \epsilon_{y_n}\}$, we obtain that

$$B_\epsilon(x) \cap Y \subset B_\epsilon(x) \cap \left(B_{\delta_{y_1}}(y_1) \cup \cdots \cup B_{\delta_{y_n}}(y_n)\right) = \varnothing$$

and thus $B_\epsilon(x) \subset X \setminus Y$. Since $x \in X \setminus Y$ was arbitrary, this means that $X \setminus Y$ is open. $\square$

Proposition 2.5.5. *Let (K, d_K) be a compact metric space, let (Y, d_Y) be any metric space, and let $f \colon K \to Y$ be continuous. Then $f(K)$ is compact.*

Proof. Let $\mathcal{U}$ be an open cover for $f(K)$. Then $\{f^{-1}(U) : U \in \mathcal{U}\}$ is an open cover for K by Corollary 2.3.10. Hence, there are $U_1, \ldots, U_n \in \mathcal{U}$ with

$$K = f^{-1}(U_1) \cup \cdots \cup f^{-1}(U_n)$$

and thus

$$f(K) \subset U_1 \cup \cdots \cup U_n.$$

This proves the claim. $\square$

Corollary 2.5.6. *Let (K, d) be a non-empty, compact metric space, and let $f \colon K \to \mathbb{R}$ be continuous. Then f attains both a minimum and a maximum on K.*

Proof. Let $M := \sup f(K)$. Since $f(K)$ is compact, it is bounded, so that $M < \infty$. For each $n \in \mathbb{N}$, there is $y_n \in f(K)$ such that $y_n > M - \frac{1}{n}$; it is clear that $M = \lim_{n \to \infty} y_n$. Since $f(K)$ is closed in $\mathbb{R}$, it follows that $M \in f(K)$. Hence, there is $x_0 \in K$ such that $f(x_0) = M$.

An analogous argument works for $\inf f(K)$. $\square$

The real line $\mathbb{R}$ has the Bolzano–Weierstraß property: every bounded sequence in $\mathbb{R}$ has a convergent subsequence. The following lemma asserts that compact metric spaces enjoy a similar property:

Lemma 2.5.7. *Let (K, d) be a compact metric space. Then every sequence in K has a convergent subsequence.*

Proof. Let $(x_n)_{n=1}^{\infty}$ be a sequence in K. Assume that $(x_n)_{n=1}^{\infty}$ has *no* convergent subsequence. This means that, for each $x \in X$ (it cannot be the limit of any subsequence of $(x_n)_{n=1}^{\infty}$!) there is $\epsilon_x > 0$ such that $B_{\epsilon_x}(x)$ contains only finitely many terms of $(x_n)_{n=1}^{\infty}$; that is, there is $n_x \in \mathbb{N}$ such that $x_n \notin B_{\epsilon_x}(x)$ for $n \geq n_x$. Since $\{B_{\epsilon_x}(x) : x \in K\}$ is an open cover for K, there are $x_1', \ldots, x_m' \in K$ with

$$K = B_{\epsilon_{x_1'}}(x_1') \cup \cdots \cup B_{\epsilon_{x_m'}}(x_m').$$

For $n \geq \max\{n_{x_1'}, \ldots, n_{x_m'}\}$, this means that

$$x_n \notin B_{\epsilon_{x_1'}}(x_1') \cup \cdots \cup B_{\epsilon_{x_m'}}(x_m') = K,$$

which is absurd. $\square$

Proposition 2.5.8. *Let (K, d) be a compact metric space. Then K is both complete and separable.*

Proof. Let $(x_n)_{n=1}^{\infty}$ be a Cauchy sequence in K. By Lemma 2.5.7, $(x_n)_{n=1}^{\infty}$ has a convergent subsequence, say $(x_{n_k})_{k=1}^{\infty}$, whose limit we denote by x. Let $\epsilon > 0$. Then there is $k_\epsilon \in \mathbb{N}$ such that $d(x_{n_k}, x) < \frac{\epsilon}{2}$ for $k \geq k_\epsilon$. Furthermore, there is $n_\epsilon \in \mathbb{N}$ with $d(x_n, x_m) < \frac{\epsilon}{2}$ for $n \geq n_\epsilon$. Choose $k_0 \geq k_\epsilon$ so large that $n_{k_0} \geq n_\epsilon$. For $n \geq n_\epsilon$, we obtain that

$$d(x_n, x) \leq d(x_n, x_{n_{k_0}}) + d(x_{n_{k_0}}, x) < \frac{\epsilon}{2} + \frac{\epsilon}{2} = \epsilon.$$

It follows that $x = \lim_{n \to \infty} x_n$.

To see that K is separable, first note that $\left\{ B_{\frac{1}{n}}(x) : x \in K \right\}$, the collection of all open balls in K of radius $\frac{1}{n}$, is an open cover for K for each $n \in \mathbb{N}$. Since K is compact, each such open cover has a finite subcover: there are, for each $n \in \mathbb{N}$, a positive integer m_n as well as $x_{1,n}, \ldots, x_{m_n, n} \in K$ such that

$$K = B_{\frac{1}{n}}(x_{1,n}) \cup \cdots \cup B_{\frac{1}{n}}(x_{m_n,n}).$$

The set $\bigcup_{n=1}^{\infty} \{x_{1,n}, \ldots, x_{m_n,n}\}$ is clearly countable. We claim that it is also dense in K. To see this, let $x \in K$, and let $\epsilon > 0$. Let $n \in \mathbb{N}$ be so large that $\frac{1}{n} < \epsilon$. Since $K = B_{\frac{1}{n}}(x_{1,n}) \cup \cdots \cup B_{\frac{1}{n}}(x_{m_n,n})$, there is $j \in \{1, \ldots, m_n\}$ such that $x \in B_{\frac{1}{n}}(x_{j,n})$ and thus $x_{j,n} \in B_\epsilon(x)$. $\square$

We now turn to two notions related to compactness.

Definition 2.5.9. *Let (X, d) be a metric space. Then:*

(a) X *is called* totally bounded *if, for each $\epsilon > 0$, there are $x_1, \dots, x_n \in X$ with*

$$X = B_\epsilon(x_1) \cup \cdots \cup B_\epsilon(x_n).$$

(b) X *is called* sequentially compact *if every sequence in X has a convergent subsequence.*

Some relations among compactness, total boundedness, and sequential compactness are straightforward. Every compact metric space is trivially totally bounded and also sequentially compact by Lemma 2.5.7. On the other hand, $(0,1)$ is easily seen to be totally bounded, but fails to be compact. The following theorem relates compactness, total boundedness, and sequential compactness in the best possible manner.

Theorem 2.5.10. *The following are equivalent for a metric space (X, d).*

(i) X *is compact.*
(ii) X *is complete and totally bounded.*
(iii) X *is sequentially compact.*

Proof. By Lemma 2.5.7, (i) $\implies$ (iii) holds.

(iii) $\implies$ (ii): The same argument as in the proof of Proposition 2.5.8 shows that X is complete. Assume that X is not totally bounded. Then there is ϵ_0 such that

$$B_{\epsilon_0}(x_1') \cup \cdots \cup B_{\epsilon_0}(x_n') \subsetneq X$$

for any choice of $x_1', \dots, x_n' \in X$. We use this to inductively construct a sequence in X that has no convergent subsequence. Let $x_1 \in X$ be arbitrary. Pick $x_2 \in X \setminus B_{\epsilon_0}(x_1)$. Then pick $x_3 \in X \setminus (B_{\epsilon_0}(x_1) \cup B_{\epsilon_0}(x_2))$. Continuing in this fashion, we obtain a sequence $(x_n)_{n=1}^\infty$ in X with

$$x_{n+1} \notin B_{\epsilon_0}(x_1) \cup \cdots \cup B_{\epsilon_0}(x_n) \qquad (n \in \mathbb{N}).$$

It is clear from this construction that

$$d(x_n, x_m) \geq \epsilon_0 \qquad (n, m \in \mathbb{N}, \, n \neq m),$$

so that no subsequence of $(x_n)_{n=1}^\infty$ can be a Cauchy sequence. This is impossible if X is sequentially compact.

(ii) $\implies$ (i): Let $\mathcal{U}$ be an open cover of X, and assume that it has no finite subcover. Since X is totally bounded, it can be covered by finitely many open balls of radius 1. Consequently, there is at least one $x_1 \in X$ such that $B_1(x_1)$ cannot be covered by finitely many sets from $\mathcal{U}$. Again by the total boundedness of X, the open ball $B_1(x_1)$ can be covered by finitely many open balls of radius $\frac{1}{2}$ (not necessarily centered at points of $B_1(x_1)$). Consequently, there is at least one $x_2 \in X$ such that $B_{\frac{1}{2}}(x_2) \cap B_1(x_1)$ cannot be covered by finitely many sets from $\mathcal{U}$. Continuing this construction, we obtain a sequence $(x_n)_{n=1}^\infty$ in X such that

$$B_{\frac{1}{n}}(x_n) \cap \cdots \cap B_{\frac{1}{2}}(x_2) \cap B_1(x_1)$$

cannot be covered by finitely many sets from $\mathcal{U}$. For $n \in \mathbb{N}$, let

$$F_n := \overline{B_{\frac{1}{n}}(x_n) \cap \cdots \cap B_{\frac{1}{2}}(x_2) \cap B_1(x_1)}.$$

Since $\operatorname{diam}(F_n) \leq \frac{2}{n} \to 0$, Cantor's intersection theorem yields that $\bigcap_{n=1}^{\infty} F_n = \{x\}$ for some $x \in X$. Let $U_0 \in \mathcal{U}$ be such that $x \in U_0$, and let $\epsilon > 0$ be such that $B_\epsilon(x) \subset U_0$. Choose $n_\epsilon \in \mathbb{N}$ such that $\frac{2}{n_\epsilon} < \epsilon$. Since $\operatorname{diam}(F_{n_\epsilon}) \leq \frac{2}{n_\epsilon}$, this means that $F_{n_\epsilon} \subset B_\epsilon(x) \subset U_0$. In particular, $\{U_0\}$ is a finite cover of $B_{\frac{1}{n_\epsilon}}(x_{n_\epsilon}) \cap \cdots \cap B_1(x_1)$, which is impossible according to our construction. $\square$

Corollary 2.5.11. *Let (X, d) be a totally bounded metric space. Then its completion is compact.*

Proof. Let $\left(\tilde{X}, \tilde{d}\right)$ be the completion of (X, d). For $r > 0$ and $x \in X$, we write $B_r(x; X)$ and $B_r\left(x; \tilde{X}\right)$ for the open balls with radius r centered at x in X and $\tilde{X}$, respectively.

Let $\epsilon > 0$. Since X is totally bounded, there are $x_1, \ldots, x_n \in X$ such that

$$X = B_{\frac{\epsilon}{2}}(x_1; X) \cup \cdots \cup B_{\frac{\epsilon}{2}}(x_n; X) \subset B_{\frac{\epsilon}{2}}\left(x_1; \tilde{X}\right) \cup \cdots \cup B_{\frac{\epsilon}{2}}\left(x_n; \tilde{X}\right).$$

Now, $\overline{B_{\frac{\epsilon}{2}}\left(x_1; \tilde{X}\right) \cup \cdots \cup B_{\frac{\epsilon}{2}}\left(x_n; \tilde{X}\right)}$ is a closed subset of $\tilde{X}$ containing X and therefore must be all of $\tilde{X}$. Since $B_{\frac{\epsilon}{2}}\left(x_j; \tilde{X}\right) \subset B_\epsilon\left(x_j; \tilde{X}\right)$ for $j = 1, \ldots, n$, we obtain that

$$\tilde{X} = B_\epsilon\left(x_1; \tilde{X}\right) \cup \cdots \cup B_\epsilon\left(x_n; \tilde{X}\right).$$

Hence, $\tilde{X}$ is also totally bounded and thus compact by Theorem 2.5.10. $\square$

The Heine–Borel theorem, which characterizes the compact subsets of $\mathbb{R}^n$, is probably familiar from several variable calculus. At the end of this section, we deduce it from Theorem 2.5.10, thus increasing our stock of compact and noncompact metric spaces.

Corollary 2.5.12 (Heine–Borel theorem). *Let $K \subset \mathbb{R}^n$. Then K is compact if and only if it is bounded and closed in $\mathbb{R}^n$.*

Proof. In view of Example 2.5.3(b) and Proposition 2.5.4(ii), the "only if" part is clear.

For the converse, first note that, since K is bounded, there is $r > 0$ such that $K \subset [-r, r]^n$. Since K is closed in $\mathbb{R}^n$ and therefore in $[-r, r]^n$, we can invoke Proposition 2.5.4(i) and suppose without loss of generality that $K = [-r, r]^n$.

As a closed subset of a complete metric space, K is clearly complete. It is therefore sufficient to show that K is totally bounded. Let $\epsilon > 0$. For $m \in \mathbb{N}$ and $j \in \{1, \ldots, m\}$, let

$$I_j := \left[-r + (j-1)\frac{2r}{m}, -r + j\frac{2r}{m} \right],$$

and note that

$$[-r, r] = \bigcup_{j=1}^{m} I_j$$

and thus

$$K = \bigcup_{(j_1,\ldots,j_n) \in \{1,\ldots,m\}^n} I_{j_1} \times \cdots \times I_{j_n}.$$

Let $(j_1, \ldots, j_n) \in \{1, \ldots, m\}^n$, and let $x, y \in I_{j_1} \times \cdots \times I_{j_n}$. The Euclidean distance of x and y can then be estimated via

$$\|x - y\| = \sqrt{\sum_{k=1}^{n} (x_k - y_k)^2} \leq \sqrt{\sum_{k=1}^{n} \left(\frac{2r}{m}\right)^2} = \frac{2r}{m}\sqrt{n}.$$

Let m be so large that $\frac{2r}{m}\sqrt{n} < \epsilon$. For $(j_1, \ldots, j_n) \in \{1, \ldots, m\}^n$, let $x_{(j_1,\ldots,j_n)} \in I_{j_1} \times \cdots \times I_{j_n}$. By the foregoing estimate, $I_{j_1} \times \cdots \times I_{j_n} \subset B_\epsilon \left(x_{(j_1,\ldots,j_n)} \right)$ holds, so that

$$K \subset \bigcup_{(j_1,\ldots,j_n) \in \{1,\ldots,m\}^n} B_\epsilon \left(x_{(j_1,\ldots,j_n)} \right).$$

Consequently, K is totally bounded and therefore compact. $\square$

Outside the realm of Euclidean n-space, the Heine–Borel theorem is no longer true, and even worse: for general metric spaces, it fails to make sense. First of all, every metric space is closed in itself, so that requiring a set to be closed depends very much on the metric space in which we are considering it. Secondly, what does it mean for a subset of a metric space to be bounded? We could, of course, define a set to be bounded if it has finite diameter, but since *every* metric is equivalent to a metric that attains its values in $[0, 1)$, and since compactness is not characterized via a particular metric, but rather through open sets, boundedness cannot be used in general metric spaces to characterize compactness.

In normed spaces, it still makes sense to speak of bounded sets as in $\mathbb{R}^n$, but the Heine–Borel theorem becomes false.

Example 2.5.13. Let $E = C([0, 1], \mathbb{F})$ be equipped with $\|\cdot\|_\infty$, and let $(f_n)_{n=1}^\infty$ be defined by

$$f_n \colon [0, 1] \to \mathbb{R}, \quad t \mapsto t^n \quad (n \in \mathbb{N}).$$

This sequence is contained in the closed unit ball $B_1[0]$ of E. If the Heine–Borel theorem is true for E, then $B_1[0]$ is compact and, consequently, $(f_n)_{n=1}^\infty$ has a convergent subsequence, say $(f_{n_k})_{k=1}^\infty$, with limit f. Since $f_n \in S_1[0]$ for $n \in \mathbb{N}$ and since $S_1[0]$ is closed, it is clear that $f \in S_1[0]$; that is, $\|f\|_\infty = 1$. On the other hand, convergence in E is uniform convergence and thus entails pointwise convergence. Hence, we have for $t \in [0, 1)$ that

$$f(t) = \lim_{k \to \infty} f_{n_k}(t) = \lim_{k \to \infty} t^{n_k} = 0.$$

Since f is continuous, this means that $f(1) = 0$ as well and thus $f \equiv 0$. This is a contradiction.

More generally, the closed unit ball of a normed space E is compact if and only if $\dim E < \infty$ (Theorem B.5).

Exercises

1. Show that a discrete metric space (X, d) is compact if and only if X is finite.
2. Let (X, d) be a metric space, and let $(x_n)_{n=1}^\infty$ be a sequence in X with limit x_0. Show that the subset $\{x_0, x_1, x_2, \ldots\}$ of X is compact.
3. Let (X, d) be a metric space, and let F and K be subspaces of X such that F is closed in X and K is compact. Show that

$$F \cap K \neq \varnothing \quad \Longleftrightarrow \quad \inf\{d(x, y) : x \in F,\ y \in K\} = 0.$$

What happens if we replace the compactness of K by the demand that it be closed in X?
4. Let $(K_1, d_1), \ldots, (K_n, d_n)$ be compact metric spaces, and let $K := K_1 \times \cdots \times K_n$ be equipped with any of the two (equivalent) metrics D_1 and D_∞ from Example 2.3.13(b). Show that K is compact.
5. More generally, let $((K_n, d_n))_{n=1}^\infty$ be a sequence of compact metric spaces, and let $K := \prod_{n=1}^\infty K_n$ be equipped with a metric d as in Example 2.1.2(h). Show that (K, d) is compact.
6. Let E be a normed space, and let $K, L \subset E$ be compact. Show that $K + L := \{x + y : x \in K,\ y \in L\}$ is also compact.
7. A subset S of a metric space (X, d) is called *relatively compact* if $\overline{S}$ is compact. Show that $S \subset X$ is relatively compact if and only if each sequence in S has a subsequence that converges in X. To what more familiar notion is relative compactness equivalent if the surrounding space X is complete?
8. Let (X, d_X) and (Y, d_Y) be metric spaces. A function $f : X \to Y$ is called *uniformly continuous* if, for each $\epsilon > 0$, there is $\delta > 0$ such that $d_Y(f(x), f(y)) < \epsilon$ whenever $d_X(x, y) < \delta$. Show that any continuous function from a compact metric space into another metric space is uniformly continuous.
9. *Lebesgue's covering lemma.* Let (K, d) be a compact metric space, and let $\mathcal{U}$ be an open cover of K. Show that there is a number $L(\mathcal{U}) > 0$ (the *Lebesgue number* of $\mathcal{U}$) such that any $\varnothing \neq S \subset K$ with $\text{diam}(S) < L(\mathcal{U})$ is contained in some $U \in \mathcal{U}$.

Remarks

Metric spaces are little more than one hundred years old: their axioms appear for the first time in Maurice Fréchet's thesis [FRÉCHET 06] from 1906. Instead of metric spaces, Fréchet speaks of classes (E), and the distance of two elements with respect to the given metric is called their *écart*, which is French for *gap*. A few years later, the German mathematician Felix Hausdorff rechristened Fréchet's classes (E) in his treatise [HAUSDORFF 14]: he called them *metrische Räume*, which translates into English literally as *metric spaces*. Most of the material from Sections 2.1, 2.2, 2.3, and 2.5 can already be found in [HAUSDORFF 14].

What we call a semimetric is usually called a *pseudometric*. However, a map p from a linear space into $[0, \infty)$ that satisfies all the axioms of a norm, except that it allows that $p(x) = 0$ for nonzero x, is called a *semi*norm, not a *pseudo*norm. This is our reason for deviating from the standard terminology, so that $p(x - y)$ for a *semi*norm p defines a *semi*metric, which is a metric if and only if p is a norm.

Bourbaki's Mittag-Leffler theorem (Theorem 2.4.14) is from "his" monumental treatise *Eléments de mathématique* [BOURBAKI 60]. The possessive pronoun is in quotation marks because Nicolas Bourbaki is not one man but the collective pseudonym of a group of French mathematicians that formed in 1935 and, from 1939 on, started publishing the aforementioned multivolume opus *Eléments de mathématique* with the goal to rebuild mathematics from scratch. Members of Nicolas Bourbaki have to leave once they reach age 50, and new members are appointed to replace the retiring ones. Hence, Nicolas Bourbaki is a truly immortal mathematician! Even though it is widely claimed (and believed), Nicolas Bourbaki was *not* the name of a French general in the Franco–Prussian war of 1871: there was a general in that war by the last name of Bourbaki, but his first names were Charles Denis. (He was offered the throne of Greece in 1862, which he turned down, and in the Franco–Prussian war, he unsuccessfully attempted suicide in order to avoid the humiliation of surrender.)

For a good reason, our Theorem 2.4.14 is somewhat less general than the result from [BOURBAKI 60]. As Jean Esterle remarks in [ESTERLE 84]:

> Incidentally, the reader interested in a French way of writing a result as clear as Corollary 2.2 [$\approx$ Theorem 2.4.14] in a form almost inaccessible to human mind is referred to the statement by Bourbaki [...].

In statement and proof of Theorem 2.4.14, we follow [DALES 78].

Baire's theorem is sometimes referred to as *Baire's category theorem* (especially in older books). The reasons for this are historical. A subset of a metric space is called *nowhere dense* if its closure has an empty interior. A subset that is a countable union of nowhere dense sets used to be called a set of the *first category* in the space, and all other subsets were said to be of the *second category*. In this terminology, Baire's theorem (or rather Corollary

2.4.17) asserts that every complete metric space is of the second category in itself. The first/second category terminology has not withstood the test of time (when mathematicians nowadays speak of categories, they mean something completely different), but the nametag *category theorem* still survives to this day.

Maurice Fréchet died in 1973, at the age of 94, decades after the concept he had introduced in his thesis had become a mathematical household item.

3

Set-Theoretic Topology

Why would one want to attempt to extend notions such as convergence and continuity to a setting even more abstract than metric spaces?

The answer is that, already at a very elementary level, one encounters phenomena that do not fit into the framework of metric spaces: pointwise convergence, for instance—the most basic notion of convergence there is for functions—cannot be described as convergence with respect to a metric (as we show in this chapter).

Convergence and continuity in the metric setting were based on a notion of "closeness" for points: two points were sufficiently close if their distance, measured through the given metric, was sufficiently small. Going beyond metric spaces and still being able to meaningfully speak of convergence and continuity therefore ought to be based on an axiomatized notion of closeness. Such an axiomatization exists (and is surprisingly simple): it lies at the heart of the concept of a topological space. (For technical reasons, we pursue a slightly different, but equivalent route.)

3.1 Topological Spaces—Definitions and Examples

A topological space is supposed to be a set that has just enough structure to meaningfully speak of continuous functions on it. In view of Corollary 2.3.10, a reasonable approach would be to axiomatize the notion of an open set:

Definition 3.1.1. *Let X be a set. A* topology *on X is a subset $\mathcal{T}$ of $\mathfrak{P}(X)$ such that:*

(a) $\varnothing, X \in \mathcal{T}$;
(b) *If $\mathcal{U} \subset \mathcal{T}$ is arbitrary, then $\bigcup \{U : U \in \mathcal{U}\}$ lies in $\mathcal{T}$;*
(c) *If $U_1, U_2 \in \mathcal{T}$, then $U_1 \cap U_2 \in \mathcal{T}$.*

The sets in $\mathcal{T}$ are called open. *A set together with a topology is called a* topological space.

We often write $(X, \mathcal{T})$ for a topological space X with topology $\mathcal{T}$; some-times, if the topology is obvious or irrelevant, we may also simply write X.

Examples 3.1.2. (a) Let (X, d) be a metric space, and let $\mathcal{T}$ denote the col-lection of all subsets of X that are open in the sense of Definition 2.2.3. By Proposition 2.2.5, $\mathcal{T}$ is indeed a topology. It is clear that $\mathcal{T}$ does not depend on the particular metric d, but only on its equivalence class: any metric on X equivalent to d yields the same topology. Topological spaces of this type are called *metrizable*.

(b) Let X be any set, and let $\mathcal{T} = \mathfrak{P}(X)$. This is just a special case of the first example: equip X with the discrete metric. Such topological spaces are called *discrete*.

(c) Let X be any set, and let $\mathcal{T} = \{\varnothing, X\}$. Such topological spaces are called *chaotic*.

(d) Let X be any set, and let $\mathcal{T}$ consist of $\varnothing$ and all subsets of X with finite complement.

(e) Let X be any set, and let $\mathcal{T}$ consist of $\varnothing$ and all subsets of X with countable complement.

(f) Let $(X, \mathcal{T})$ be a topological space, and let $Y \subset X$. The *relative topology* on Y (or the *topology inherited from X*) is the collection

$$\mathcal{T}|_Y := \{Y \cap U : U \in \mathcal{T}\}$$

of subsets of Y. It is clearly a topology on Y. The space $(Y, \mathcal{T}|_Y)$ is then called a *subspace* of X.

Is every topological space metrizable? Of course not, and here is why.

Definition 3.1.3. *A topological space $(X, \mathcal{T})$ is called* Hausdorff *if, for any $x, y \in X$ with $x \neq y$, there are sets $U, V \in \mathcal{T}$ with $x \in U$, $y \in V$, and $U \cap V = \varnothing$.*

Informally, Definition 3.1.3 is often expressed as, "In a Hausdorff space, points can be separated by open sets."

Examples 3.1.4. (a) Let (X, d) be a metric space, and let $x, y \in X$ be such that $x \neq y$. It follows that $\epsilon := \frac{1}{2} d(x, y) > 0$. Let $U := B_\epsilon(x)$ and let $V := B_\epsilon(y)$. It follows that $U \cap V = \varnothing$, so that X is Hausdorff.

(b) If X is any set with more than one element, then X equipped with the chaotic topology is not Hausdorff (and therefore not metrizable).

(c) Let X be an infinite set equipped with the topology from Example 3.1.2(d), and let $x, y \in X$ be such that $x \neq y$. Assume that X is Hausdorff. Then there are open sets U and V of X with $x \in U$, $y \in V$, and $U \cap V = \varnothing$. This, however, entails that $X = (X \setminus U) \cup (X \setminus V)$ is finite, which is a contradiction.

(d) Similarly, if X is any uncountable set equipped with the topology from Example 3.1.2(e), the resulting topological space fails to be Hausdorff (see Exercise 1 below).

We soon encounter Hausdorff spaces that nevertheless fail to be metrizable.

With a notion of open sets at hand, we can define, of course, what a closed subset of a topological space is supposed to be:

Definition 3.1.5. *Let $(X, \mathcal{T})$ be a topological space. A subset F of X is called closed if $X \setminus F$ is open.*

As for metric spaces, we have the following.

Proposition 3.1.6. *Let $(X, \mathcal{T})$ be a topological space. Then:*

(i) $\varnothing$ *and X are closed.*
(ii) *If $\mathcal{F}$ is a family of closed subsets of X, then $\bigcap\{F : F \in \mathcal{F}\}$ is closed.*
(iii) *If F_1 and F_2 are closed subsets of X, then $F_1 \cup F_2$ is closed.*

Of course, one can also define a topology on a given set by declaring certain sets as closed, then checking that these sets satisfy Proposition 3.1.6(i) through (iii), and defining their complements as open; it is clear that this approach is equivalent to Definition 3.1.1.

Example 3.1.7. Let R be a commutative ring with identity. Recall that a proper ideal $\mathfrak{p}$ of R (i.e., $\mathfrak{p} \subsetneq R$) is called *prime* if $ab \in \mathfrak{p}$ implies that $a \in \mathfrak{p}$ or $b \in \mathfrak{p}$. For example, let $R = \mathbb{Z}$; then every ideal of $\mathbb{Z}$ is of the form $n\mathbb{Z} := \{nm : m \in \mathbb{Z}\}$ with $n \in \mathbb{N}_0$, and it is a prime ideal of $\mathbb{Z}$ if and only if n is zero or a prime number. Let

$$\text{Spec}(R) := \{\mathfrak{p} : \mathfrak{p} \text{ is a prime ideal of } R\}.$$

We now define a topology on $\text{Spec}(R)$ by declaring certain subsets of $\text{Spec}(R)$ as closed.

For any ideal I of R, let

$$V(I) := \{\mathfrak{p} \in \text{Spec}(R) : I \subset \mathfrak{p}\},$$

so that, in particular, $\varnothing = V(R)$ and $\text{Spec}(R) = V(\{0\})$. Let $\mathcal{I}$ be a family of ideals of R, and let $\sum\{I : I \in \mathcal{I}\}$ be the set of all finite sums $\sum_{j=1}^{n} a_j$ such that there are $I_1, \ldots, I_n \in \mathcal{I}$ with $a_j \in I_j$ for $j = 1, \ldots, n$. It is clear that $\sum\{I : I \in \mathcal{I}\}$ is again an ideal of R, and it is easy to see that

$$\bigcap\{V(I) : I \in \mathcal{I}\} = V\left(\sum\{I : I \in \mathcal{I}\}\right).$$

Let I_1 and I_2 be ideals of R, and let I be the ideal of R generated by the set $\{ab : a \in I_1, b \in I_2\}$. It is easy to see that I consists precisely of those elements of R that are of the form $\sum_{j=1}^{n} a_j b_j$ with $a_1, \ldots, a_n \in I_1$ and $b_1, \ldots, b_n \in I_2$. We claim that $V(I_1) \cup V(I_2) = V(I)$. If $\mathfrak{p}$ is a prime ideal containing both I_1 or I_2, it is clear that $I \subset \mathfrak{p}$. Consequently, $V(I_1) \cup V(I_2) \subset V(I)$ holds. Conversely, let $\mathfrak{p} \in V(I)$, and, without loss of generality, suppose that $\mathfrak{p} \notin V(I_1)$, so that there is $a \in I_1$ with $a \notin \mathfrak{p}$. Since $\mathfrak{p} \in V(I)$, it follows that

$$\{ab : b \in I_2\} \subset I \subset \mathfrak{p}$$

and therefore, because $\mathfrak{p}$ is a prime ideal, that $I_2 \subset \mathfrak{p}$; that is, $\mathfrak{p} \in V(I_2)$. All in all, we obtain $V(I_1) \cup V(I_2) = V(I)$ as claimed.

The sets of the form $V(I)$, where I is an ideal of R, are thus the closed sets of a topology on $\operatorname{Spec}(R)$. This topology is called the *Zariski topology*.

With a notion of openness, we can define neighborhoods in topological spaces.

Definition 3.1.8. *Let $(X,\mathcal{T})$ be a topological space, and let $x \in X$. A subset N of X is called a* neighborhood *of x if there is an open subset U of X with $x \in U \subset N$. The collection of all neighborhoods of x is denoted by $\mathcal{N}_x$.*

The following proposition is proven as for metric spaces.

Proposition 3.1.9. *Let $(X,\mathcal{T})$ be a topological space, and let $x \in X$. Then:*

(i) *If $N \in \mathcal{N}_x$ and $M \supset N$, then $M \in \mathcal{N}_x$.*
(ii) *If $N_1, N_2 \in \mathcal{N}_x$, then $N_1 \cap N_2 \in \mathcal{N}_x$.*

Moreover, a subset U of X is open if and only if $U \in \mathcal{N}_y$ for each $y \in U$.

The following is of interest because it shows that, instead of through axiomatizing the notion of openness, a topology can also be defined via an axiomatized notion of neighborhood.

Theorem 3.1.10. *Let X be a set, and let, for each $x \in X$, there be $\varnothing \neq \mathfrak{N}_x \subset \mathfrak{P}(X)$ such that:*

(a) *$x \in N$ for each $N \in \mathfrak{N}_x$;*
(b) *If $N \in \mathfrak{N}_x$ and $M \supset N$, then $M \in \mathfrak{N}_x$;*
(c) *If $N_1, N_2 \in \mathfrak{N}_x$, then $N_1 \cap N_2 \in \mathfrak{N}_x$;*
(d) *For each $N \in \mathfrak{N}_x$ there is $U \in \mathfrak{N}_x$ such that $U \subset N$ and $U \in \mathfrak{N}_y$ for all $y \in U$.*

Let $\mathcal{T}$ be the collection of all subsets U of X with $U \in \mathfrak{N}_y$ for each $y \in U$. Then $\mathcal{T}$ is the unique topology on X such that $\mathfrak{N}_x = \mathcal{N}_x$ for each $x \in X$.

Proof. Trivially, $\varnothing$ and X are in $\mathcal{T}$.

Let $\mathcal{U} \subset \mathcal{T}$, and let $y \in \bigcup\{U : U \in \mathcal{U}\}$. It follows that there is $U_0 \in \mathcal{U}$ with $y \in U_0$; that is, $U_0 \in \mathfrak{N}_y$. By (b), this means that $\bigcup\{U : U \in \mathcal{U}\} \in \mathfrak{N}_y$ as well. Since y was arbitrary, this means that $\bigcup\{U : U \in \mathcal{U}\} \in \mathcal{T}$, too.

Let $U_1, U_2 \in \mathcal{T}$, and let $y \in U_1 \cap U_2$; that is, $U_1, U_2 \in \mathfrak{N}_y$. From (c), it follows that $U_1 \cap U_2 \in \mathfrak{N}_y$. Again, since y was arbitrary, $U_1 \cap U_2 \in \mathcal{T}$ follows.

All in all, $\mathcal{T}$ is a topology, so that it makes sense to speak of $\mathcal{N}_x$ for $x \in X$. (Note that no use has been made so far of (d).)

Let $x \in X$, and let $N \in \mathfrak{N}_x$. By (d), there is $U \in \mathfrak{N}_x$ such that $U \in \mathfrak{N}_y$ for each $y \in U$. By the definition of $\mathcal{T}$, this means that U is open, and since $x \in U \subset N$, we have $N \in \mathcal{N}_x$. Conversely, if $N \in \mathcal{N}_x$, there is $U \in \mathcal{T}$ with

$x \in U \subset N$. By the definition of $\mathcal{T}$, we have $U \in \mathfrak{N}_x$, so that $N \in \mathfrak{N}_x$ by (b). Hence, $\mathfrak{N}_x = \mathcal{N}_x$ holds for all $x \in X$.

From the "moreover" statement of Proposition 3.1.9, it is clear that $\mathcal{T}$ is uniquely determined by this property. $\square$

Example 3.1.11. Let $S \neq \varnothing$ be a set, let (Y, d) be a metric space, and let $F(S, Y)$ denote the set of all functions from S to Y. Let $\varnothing \neq \mathcal{C} \subset \mathfrak{P}(S)$ be closed under finite unions. For $f \in F(S, Y)$, $C \in \mathcal{C}$, and $\epsilon > 0$, let

$$N_{f,C,\epsilon} := \left\{ g \in F(S, Y) : \sup_{x \in C} d(f(x), g(x)) < \epsilon \right\},$$

and, for $f \in F(S, Y)$, let

$$\mathfrak{N}_f := \{ N \subset F(S, Y) : N \supset N_{f,C,\epsilon} \text{ for some } C \in \mathcal{C} \text{ and } \epsilon > 0 \}.$$

We claim that $\mathfrak{N}_f$ satisfies conditions (a) through (d) of Theorem 3.1.10. Trivially, (a) and (b) are satisfied, and since

$$N_{f,C_1,\epsilon_1} \cap N_{f,C_2,\epsilon_2} \supset N_{f,C_1 \cup C_2, \min\{\epsilon_1,\epsilon_2\}}$$

for $C_1, C_2 \in \mathcal{C}$ and $\epsilon_1, \epsilon_2 > 0$, it follows that (c) holds as well. To see that (d) is also true, let $N \in \mathfrak{N}_f$, so that there are $C \in \mathcal{C}$ and $\epsilon > 0$ with $U := N_{f,C,\epsilon} \subset N$. Let $g \in U$, so that $\sup_{x \in C} d(f(x), g(x)) < \epsilon$, and let $\delta := \epsilon - \sup_{x \in C} d(f(x), g(x)) > 0$. It is routinely seen that $N_{g,C,\delta} \subset U$, so that $U \in \mathfrak{N}_g$. Since $g \in U$ was arbitrary, it follows that $\mathfrak{N}_f$ also satisfies Theorem 3.1.10(d).

By Theorem 3.1.10, we therefore have a unique topology, which we denote by $\mathcal{T}_{\mathcal{C}}$, on $F(S, Y)$ with $\mathfrak{N}_f = \mathcal{N}_f$ for each $f \in F(S, Y)$. (The definition of $\mathcal{T}_{\mathcal{C}}$ may seem bewildering, but we show in the next section that such topologies can be used to capture well-known phenomena in analysis, such as pointwise and uniform convergence.)

In Example 3.1.11, we constructed a system of neighborhoods at each point of the space $F(S, Y)$ by defining those neighborhoods as containing certain, more basic sets. This may serve as motivation for the following definition.

Definition 3.1.12. *Let $(X, \mathcal{T})$ be a topological space, and let $x \in X$. A base for $\mathcal{N}_x$ is a subset $\mathcal{B}_x$ of $\mathcal{N}_x$ such that, for each $N \in \mathcal{N}_x$, there is $B \in \mathcal{B}_x$ with $B \subset N$. Neighborhoods in $\mathcal{B}_x$ are called* basic.

Examples 3.1.13. (a) In Example 3.1.11, for given $f \in F(S, Y)$, the sets of the form $N_{f,C,\epsilon}$ with $C \in \mathcal{C}$ and $\epsilon > 0$ form a base for $\mathcal{N}_f$.
(b) Let $(X, \mathcal{T})$ be any topological space, and let $x \in X$. Then $\{ U \in \mathcal{N}_x : U \text{ is open} \}$ is a base for $\mathcal{N}_x$.
(c) Let (X, d) be a metric space, and let $x \in X$. Then both $\{ B_\epsilon(x) : \epsilon > 0 \}$ and $\left\{ B_{\frac{1}{n}}(x) : n \in \mathbb{N} \right\}$ are bases for $\mathcal{N}_x$.

With the notions of a base of neighborhoods at hand, we can now exhibit topological spaces that are Hausdorff, but not metrizable.

Definition 3.1.14. *A topological space* $(X, \mathcal{T})$ *is called* first countable *if, for each* $x \in X$, *there is a countable base for* $\mathcal{N}_x$.

Examples 3.1.15. (a) In view of Example 3.1.13(c), every metrizable topological space is first countable.

(b) Let S be an uncountable set, let Y be any metric space with more than one point, and let $\mathcal{F} := \{F \subset S : F \text{ is finite}\}$. We claim that the topological space $(F(S, Y), \mathcal{T}_{\mathcal{F}})$ from Example 3.1.11 is not first countable—and thus not metrizable—but nevertheless Hausdorff.

Let $f, g \in F(S, Y)$ such that $f \neq g$. Hence, there is $x \in S$ with $f(x) \neq g(x)$. Let $\epsilon := \frac{1}{2} d(f(x), g(x))$. It follows that

$$N_{f, \{x\}, \epsilon} \cap N_{g, \{x\}, \epsilon} = \varnothing,$$

so that $(F(S, Y), \mathcal{T}_{\mathcal{F}})$ is Hausdorff.

Let $f \in F(S, Y)$, and assume that $\mathcal{N}_f$ has a countable base, $\{B_1, B_2, \ldots\}$ say. We first claim that $\bigcap_{n=1}^{\infty} B_n = \{f\}$. To see this, assume towards a contradiction that there is $g \neq f$ in $\bigcap_{n=1}^{\infty} B_n$. Since $(F(S, Y), \mathcal{T}_{\mathcal{F}})$ is Hausdorff, there is $N \in \mathcal{N}_f$ with $g \notin N$, and from the definition of a base for $\mathcal{N}_f$, we obtain that $g \notin B_m$ for some $m \in \mathbb{N}$: this contradicts our assumption. From the definition of $\mathcal{N}_f$, it follows immediately that, for each $n \in \mathbb{N}$, there are $F_n \in \mathcal{F}$ and $\epsilon_n > 0$ such that $N_{f, F_n, \epsilon_n} \subset B_n$. It follows that

$$\{f\} \subset \bigcap_{n=1}^{\infty} N_{f, F_n, \epsilon_n} \subset \bigcap_{n=1}^{\infty} B_n = \{f\}.$$

Define $g \colon S \to Y$ as follows,

$$g(x) := \begin{cases} f(x), & \text{if } x \in \bigcup_{n=1}^{\infty} F_n, \\ \text{some } y \neq f(x), & \text{otherwise}; \end{cases}$$

this is possible because Y has more than one point. Since S is uncountable, $S \supsetneq \bigcup_{n=1}^{\infty} F_n$ must hold, so that $g \neq f$. It is clear, however, that $g \in \bigcap_{n=1}^{\infty} N_{f, F_n, \epsilon_n}$, which is impossible. Consequently, $(F(S, Y), \mathcal{T}_{\mathcal{F}})$ is not first countable and therefore not metrizable.

Since, in general topological spaces, a notion of closed sets exists, we can define the closure of an arbitrary subset.

Definition 3.1.16. *Let* $(X, \mathcal{T})$ *be a topological space. For each* $S \subset X$, *the* closure *of* S *is defined as*

$$\overline{S} := \bigcap \{F : F \subset X \text{ is closed and contains } S\}.$$

Example 3.1.17. Let R be a commutative ring with identity, and let $\mathrm{Spec}(R)$ be equipped with the Zariski topology. Let $S \subset \mathrm{Spec}(R)$. We claim that

$$\overline{S} = V\left(\bigcap\{\mathfrak{p} : \mathfrak{p} \in S\}\right).$$

Since $\bigcap\{\mathfrak{p} : \mathfrak{p} \in S\}$ is an ideal of R, it follows that $V\left(\bigcap\{\mathfrak{p} : \mathfrak{p} \in S\}\right)$ is closed (and trivially contains S), so that

$$\overline{S} \subset V\left(\bigcap\{\mathfrak{p} : \mathfrak{p} \in S\}\right).$$

Let I be an ideal of R with $\overline{S} = V(I)$. It follows that $I \subset \mathfrak{p}$ for each $\mathfrak{p} \in S$ and thus $I \subset \bigcap\{\mathfrak{p} : \mathfrak{p} \in S\}$. This, in turn, yields that

$$V\left(\bigcap\{\mathfrak{p} : \mathfrak{p} \in S\}\right) \subset V(I) = \overline{S},$$

so that equality holds.

As for metric spaces, we have the following proposition.

Proposition 3.1.18. *Let $(X, \mathcal{T})$ be a topological space, and let $S \subset X$. Then we have*
$$\overline{S} = \{x \in X : N \cap S \neq \varnothing \text{ for all } N \in \mathcal{N}_x\}.$$

The proof is an almost verbatim copy of that of Proposition 2.2.13.

As with the notion of neighborhood, that of closure can be used to define a topology on a given set.

Definition 3.1.19. *Let X be a set. A* Kuratowski closure operation *is a map* $\mathrm{cl} \colon \mathfrak{P}(X) \to \mathfrak{P}(X)$ *satisfying*

(a) $\mathrm{cl}(\varnothing) = \varnothing$,
(b) $S \subset \mathrm{cl}(S)$ *for all* $S \subset X$,
(c) $\mathrm{cl}(\mathrm{cl}(S)) = \mathrm{cl}(S)$ *for all* $S \subset X$, *and*
(d) $\mathrm{cl}(S \cup T) = \mathrm{cl}(S) \cup \mathrm{cl}(T)$ *for all* $S, T \subset X$.

It is immediate that taking the closure in a topological space is a Kuratowski closure operation: (a), (b), and (c) hold trivially, and (d) is easy to verify (see Exercise 2 below).

Theorem 3.1.20. *Let X be a set equipped with a Kuratowski closure operation* cl. *Then those subsets F of X such that $\mathrm{cl}(F) = F$ form the closed subsets of a unique topology $\mathcal{T}$ on X.*

Proof. Set
$$\mathcal{T} := \{U \subset X : \mathrm{cl}(X \setminus U) = X \setminus U\}.$$
We show that $\mathcal{T}$ is a topology on X such that $\overline{S} = \mathrm{cl}(S)$ for each $S \subset X$.

We refer to a set $F \subset X$ as $\mathcal{T}$-closed if its complement in X is in $\mathcal{T}$; that is, $\mathrm{cl}(F) = F$. We check that the family of all $\mathcal{T}$-closed subsets of X satisfies properties (i), (ii), and (iii) of Proposition 3.1.6.

By Definition 3.1.19(a) and (b), $\varnothing$ and X are $\mathcal{T}$-closed, and by (d), the union of any two $\mathcal{T}$-closed sets is $\mathcal{T}$-closed again.

For $S, T \subset X$ with $S \subset T$, observe that, by (d),

$$\mathrm{cl}(S) \subset \mathrm{cl}(S) \cup \mathrm{cl}(T) = \mathrm{cl}(S \cup T) = \mathrm{cl}(T).$$

Let $\mathcal{F}$ be a family of $\mathcal{T}$-closed sets, and let $F_0 \in \mathcal{F}$. Then we have, in view of the foregoing, that

$$\mathrm{cl}\left(\bigcap\{F : F \in \mathcal{F}\}\right) \subset \mathrm{cl}(F_0) = F_0.$$

Since $F_0 \in \mathcal{F}$ was arbitrary, we obtain

$$\mathrm{cl}\left(\bigcap\{F : F \in \mathcal{F}\}\right) \subset \bigcap\{F : F \in \mathcal{F}\}.$$

The converse inclusion holds by (b), so that $\bigcap\{F : F \in \mathcal{F}\}$ is $\mathcal{T}$-closed.

All in all, $\mathcal{T}$ is a topology on X.

Let $S \subset X$. By (b), we have $S \subset \mathrm{cl}(S)$ and thus $\overline{S} \subset \mathrm{cl}(S)$ because $\mathrm{cl}(S)$ is $\mathcal{T}$-closed by (c). On the other hand, $S \subset \overline{S}$ implies that

$$\mathrm{cl}(S) \subset \mathrm{cl}\left(\overline{S}\right) = \overline{S},$$

so that in the end $\overline{S} = \mathrm{cl}(S)$. $\square$

With the definition of closure comes that of density.

Definition 3.1.21. *Let $(X, \mathcal{T})$ be a topological space. Then $D \subset X$ is said to be* dense *in X if $\overline{D} = X$.*

Example 3.1.22. Let R be a commutative ring with identity, and let $\mathfrak{p} \in \mathrm{Spec}(R)$. By Example 3.1.17, we have

$$\overline{\{\mathfrak{p}\}} = \{\mathfrak{q} \in \mathrm{Spec}(R) : \mathfrak{p} \subset \mathfrak{q}\}.$$

Suppose now that R is an *integral domain*; that is, $ab = 0$ implies $a = 0$ or $b = 0$ (this is the same as saying that the zero ideal (0) is prime). For example, $\mathbb{Z}$ and any field are integral domains whereas $\mathbb{Z}/4\mathbb{Z}$ isn't. Then we have

$$\overline{\{(0)\}} = \{\mathfrak{q} \in \mathrm{Spec}(R) : (0) \subset \mathfrak{q}\} = \mathrm{Spec}(R);$$

that is, the singleton subset $\{(0)\}$ is dense in $\mathrm{Spec}(R)$.

This example shows that general topological spaces can display somewhat bizarre phenomena that cannot occur in metric spaces: a singleton subset is dense in a metric space if and only if the whole space itself is a singleton set. Example 3.1.22 is not contrived in any way: the Zariski spectra of commutative rings are important objects in commutative algebra and algebraic geometry.

With the definition of density at hand, separability can be defined.

Definition 3.1.23. *A topological space is called* separable *if it has a dense countable subset.*

One might guess that, in analogy with the situation for metric spaces, a subspace of a separable topological space is again separable, but as we show, this is not true. We use this as an excuse to introduce yet another definition (actually two definitions).

Definition 3.1.24. *Let* (X, T) *be a topological space. Then:*

(a) *A* base *for* T *is a collection* $\mathcal{B}$ *of open sets such that each open set is a union of sets in* $\mathcal{B}$;
(b) *A* subbase *for* T *is a collection* $\mathcal{S}$ *of open sets such that the collection of all finite intersections of sets in* $\mathcal{S}$ *is a base for* T.

Examples 3.1.25. (a) Let (X, d) be a metric space. Then $\{B_r(x) : x \in X, r > 0\}$ is a base for the topology induced by d.
(b) Let R be a commutative ring with identity. For $a \in R$, let

$$V(a) := \{\mathfrak{p} \in \mathrm{Spec}(R) : a \in \mathfrak{p}\}.$$

It is clear that $V(a)$ equals $V(aR)$ and thus is closed for each $a \in R$. Let $F \subset \mathrm{Spec}(R)$ be closed; that is, $F = V(I)$ for some ideal I of R. It follows that

$$F = V(I) = \bigcap\{V(a) : a \in I\}.$$

Consequently, $\{\mathrm{Spec}(R) \setminus V(a) : a \in R\}$ is a base for the Zariski topology.

Let X be any set, and suppose that $\mathcal{B} \subset \mathfrak{P}(X)$ has the property that, for any $B_1, B_2 \in \mathcal{B}$, their intersection $B_1 \cap B_2$ is empty or belongs to $\mathcal{B}$ again. Then, as is straightforward to verify, the collection of all unions of sets from $\mathcal{B}$ is a topology on X having $\mathcal{B}$ as base. More generally, if $\mathcal{S} \subset \mathfrak{P}(X)$ is arbitrary, then the collection of all unions of finite intersections of sets from $\mathcal{S}$ is a topology on X with $\mathcal{S}$ as subbase.

Example 3.1.26. Let $X = \mathbb{R}^2$. For $a, b \in \mathbb{R}$, let

$$B_{a,b} := \{(x, y) \in \mathbb{R}^2 : x \geq a, \, y \geq b\}.$$

Then $\mathcal{B} := \{B_{a,b} : a, b \in \mathbb{R}\}$ is stable under finite intersections. Hence, by the preceding remark, there is a (necessarily unique) topology T on X with $\mathcal{B}$ as base. We first claim that (X, T) is separable. We show that $C := \{(n, n) : n \in \mathbb{N}\}$ is dense in X. Assume that C is not dense in X. Then $\varnothing \neq U := X \setminus \overline{C}$ is open and thus contains a set of the form $B_{a,b}$ with $a, b \in \mathbb{R}$. However, if $n \geq \max\{a, b\}$, it is clear that $(n, n) \in C \cap B_{a,b}$, which yields a contradiction.

Let $Y := \{(x, -x) : x \in \mathbb{R}\}$. For any $a \in \mathbb{R}$, we have

$$Y \cap B_{a,-a} = \{(a, -a)\}.$$

Hence, all singleton subsets of Y are open with respect to $\mathcal{T}|_Y$, so that the topological space $(Y, \mathcal{T}|_Y)$ must be discrete. Clearly, Y is uncountable, and that's impossible if it were separable.

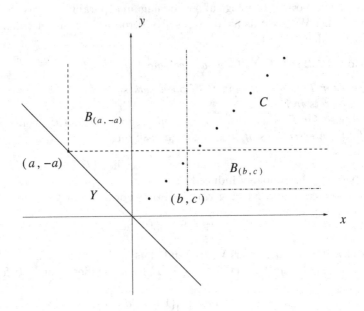

Fig. 3.1: A separable space with a nonseparable subspace

We finish this section with a brief discussion of the boundary and the interior of subsets of topological spaces: the definitions and results very much parallel those for metric spaces, and the corresponding proofs carry over.

Definition 3.1.27. *Let* $(X, \mathcal{T})$ *be a topological space, and let* $S \subset X$. *Then the* boundary *of* S *is defined as*

$$\partial S := \{x \in X : N \cap S \neq \varnothing \text{ and } N \cap (X \setminus S) \neq \varnothing \text{ for all } N \in \mathcal{N}_x\}.$$

Proposition 3.1.28. *Let* $(X, \mathcal{T})$ *be a topological space, and let* $S \subset X$. *Then:*

(i) $\partial S = \partial(X \setminus S)$;
(ii) ∂S *is closed;*
(iii) $\overline{S} = S \cup \partial S$.

Definition 3.1.29. *Let* $(X, \mathcal{T})$ *be a topological space. For each* $S \subset X$, *the* interior *of* S *is defined as*

$$\overset{\circ}{S} := \bigcup \{U : U \subset X \text{ is open and contained in } S\}.$$

Proposition 3.1.30. *Let $(X, \mathcal{T})$ be a topological space, and let $S \subset X$. Then we have:*

$$\overset{\circ}{S} = \{x \in X : S \in \mathcal{N}_x\} = S \setminus \partial S.$$

Exercises

1. Let $(X, \mathcal{T})$ be the topological space from Example 3.1.2(e); that is, $U \subset X$ is open if and only if U is empty or has a countable complement. Show that X is countable if and only if $(X, \mathcal{T})$ is discrete and if and only if $(X, \mathcal{T})$ is Hausdorff.

2. Let $(X, \mathcal{T})$ be a topological space, and let $S, T \subset X$. Show that $\overline{S \cup T} = \overline{S} \cup \overline{T}$ and $\overline{S} \setminus \overline{T} \subset \overline{S \setminus T}$.

3. Let R be a commutative ring with identity, and let $\mathfrak{p} \in \mathrm{Spec}(R)$. Show that $\{\mathfrak{p}\}$ is closed in $\mathrm{Spec}(R)$ if and only if $\mathfrak{p}$ is a maximal ideal of R.

4. A topological space $(X, \mathcal{T})$ is called *second countable* if $\mathcal{T}$ has a countable base. Show that:
   (a) Every second countable space is first countable;
   (b) Every separable metric space is second countable.

5. Let $(X, \mathcal{T})$ be the separable space from Example 3.1.26. Show that X is first countable, but not second countable. (This space is not Hausdorff; in Example 3.5.14 below, a separable Hausdorff space—easily seen to be first countable—is discussed that fails to be second countable.)

6. *A topological proof for the infinitude of primes.* For any $a \in \mathbb{Z}$ and $b \in \mathbb{N}$, let

$$N_{a,b} := \{a + nb : n \in \mathbb{Z}\}.$$

For $a \in \mathbb{Z}$, let $\mathfrak{N}_a$ consist of those sets N such that there is $b \in \mathbb{N}$ with $N_{a,b} \subset N$.
   (a) Show that, for each $a \in \mathbb{Z}$, the system $\mathfrak{N}_a$ satisfies the hypotheses of Theorem 3.1.10, so that there is a unique topology $\mathcal{T}$ on $\mathbb{Z}$ such that $\mathfrak{N}_a = \mathcal{N}_a$ for each $a \in \mathbb{Z}$.
   (b) Argue that any open subset of $(\mathbb{Z}, \mathcal{T})$ is either empty or infinite.
   (c) Show that, for any $a \in \mathbb{Z}$ and $b \in \mathbb{N}$, the set $N_{a,b}$ is both open and closed.
   (d) Argue that

$$\mathbb{Z} \setminus \{-1, 1\} = \bigcup \{N_{0,p} : p \text{ is a prime number}\}.$$

   (e) Conclude that there are infinitely many prime numbers.

7. Let $(X, \mathcal{T})$ be a topological space, and let $S \subset X$. Show that $X \setminus \overset{\circ}{S} = \overline{X \setminus S}$ and $X \setminus \overline{S} = (X \setminus S)^{\circ}$.

8. *Quotient spaces.* Let $(X, \mathcal{T})$ be a topological space, and let $\approx$ be an equivalence relation on X. For $x \in X$, let $[x]$ denote its equivalence class with respect to $\approx$, and let $X/\approx$ denote the collection of all $[x]$ with $x \in X$. Show that the collection of all subsets U of $X/\approx$ such that $\{x \in X : [x] \in U\} \in \mathcal{T}$ is a topology on $X/\approx$. (This topology is called the *quotient topology* on $X/\approx$, and the resulting topological space is the *quotient space* of X with respect to $\approx$.)

3.2 Continuity and Convergence of Nets

One can, of course, define convergence for sequences in topological spaces as for metric spaces.

Definition 3.2.1. *Let $(X, \mathcal{T})$ be a topological space. A sequence $(x_n)_{n=1}^{\infty}$ in X is said to* converge *to $x \in X$ if, for each $N \in \mathcal{N}_x$, there is $n_N \in \mathbb{N}$ such that $x_n \in N$ for all $n \geq n_N$.*

This definition is perfectly fine, but if one attempts to prove analogues of results for convergent sequences in metric spaces, problems show up. Proposition 2.3.4, for example, is no longer true in general topological spaces.

Example 3.2.2. Let X be an uncountable set equipped with the topology of Example 3.1.2(e); that is, the open sets are $\varnothing$ and those with a countable complement. Fix a point $x_0 \in X$. Then $X \setminus \{x_0\}$ is not closed, so that $\overline{X \setminus \{x_0\}} = X$ must hold. Let $(x_n)_{n=1}^{\infty}$ be a sequence in $X \setminus \{x_0\}$, and let $U := X \setminus \{x_1, x_2, \ldots\}$. Due to the nature of our topology, U is open and thus is a neighborhood of x_0. However, $x_n \notin U$ for all $n \in \mathbb{N}$ by definition, so that $(x_n)_{n=1}^{\infty}$ cannot converge to x_0.

A less contrived example for the failure of Proposition 2.3.4 in general topological spaces is given in Exercise 11 below.

So, how are we going to define continuity on arbitrary topological spaces? Of course, we could try it via sequences as for metric spaces, but in view of Example 3.2.2, we are likely to run into unexpected difficulties. Of the four equivalent conditions of Theorem 2.3.7, the fourth one doesn't make any explicit reference to a metric. We thus use it as the definition of continuity.

Definition 3.2.3. *Let $(X, \mathcal{T}_X)$ and $(Y, \mathcal{T}_Y)$ be topological spaces. Then $f : X \to Y$ is said to be* continuous *at $x_0 \in X$ if $f^{-1}(N) \in \mathcal{N}_{x_0}$ for each $N \in \mathcal{N}_{f(x_0)}$. If f is continuous at every point of X, we simply call f continuous.*

As for metric spaces (Corollary 2.3.10), we have (with an identical proof):

Proposition 3.2.4. *Let $(X, \mathcal{T}_X)$ and $(Y, \mathcal{T}_Y)$ be topological spaces. Then the following are equivalent for $f : X \to Y$.*

(i) *f is continuous.*
(ii) *$f^{-1}(U)$ is open in X for each open subset U of Y.*
(iii) *$f^{-1}(F)$ is closed in X for each closed subset F of Y.*

Examples 3.2.5. (a) Let $(X, \mathcal{T}_X)$ and $(Y, \mathcal{T}_Y)$ be topological spaces such that X is discrete. Then every function $f : X \to Y$ is continuous.
(b) Let $(X, \mathcal{T}_X)$ be a chaotic topological space (i.e., $\mathcal{T}_X = \{\varnothing, X\}$), let $(Y, \mathcal{T}_Y)$ be a Hausdorff space, and let $f : X \to Y$ be continuous. We claim that f has to be constant. Otherwise, there are $x, y \in X$ such that $f(x) \neq f(y)$. Choose $U, V \in \mathcal{T}_Y$ such that $f(x) \in U$, $f(y) \in V$, and $U \cap V = \varnothing$. Since f is continuous, $f^{-1}(U)$ is open in X and nonempty because it contains

x. Due to the definition of a chaotic topology, $f^{-1}(U)$ must equal X. This, in turn, implies that $y \in f^{-1}(U)$ and therefore $f(y) \in U$, which is a contradiction. (The demand that $(Y, \mathcal{T}_Y)$ be Hausdorff cannot be dropped; if Y is also chaotic, every map from X to Y is continuous.)

These two examples show drastically, how the continuous functions between topological spaces depend on the topologies with which those spaces are equipped. Sometimes, we want particular maps between sets to be continuous and will adjust the topologies accordingly.

Definition 3.2.6. *Let X be a nonempty set, and let $\mathcal{T}_1$ and $\mathcal{T}_2$ be topologies on X. We say that $\mathcal{T}_1$ and $\mathcal{T}_2$ are* comparable *if $\mathcal{T}_1 \subset \mathcal{T}_2$ or $\mathcal{T}_2 \subset \mathcal{T}_1$. If $\mathcal{T}_1 \subset \mathcal{T}_2$, we say that $\mathcal{T}_1$ is* coarser *than $\mathcal{T}_2$ or, equivalently, that $\mathcal{T}_2$ is* finer *than $\mathcal{T}_1$.*

Clearly, $\mathcal{T}_1$ is finer than $\mathcal{T}_2$ if and only if $\mathrm{id} \colon (X, \mathcal{T}_1) \to (X, \mathcal{T}_2)$ is continuous.

Proposition 3.2.7. *Let X be a set, let $((Y_i, \mathcal{T}_i))_{i \in \mathbb{I}}$ be a family of topological spaces, and let $f_i \colon X \to Y_i$ be a function for each $i \in \mathbb{I}$. Then there is a coarsest topology on X such that each of the maps f_i is continuous. The collection $\{f_i^{-1}(U) : i \in \mathbb{I}, U \in \mathcal{T}_i\}$ is a subbase for this topology.*

Proof. Let $\mathcal{S} := \{f_i^{-1}(U) : i \in \mathbb{I}, U \in \mathcal{T}_i\}$, and let $\mathcal{T}$ be the collection of all unions of finite intersections of sets from $\mathcal{S}$. Then $\mathcal{T}$ is a topology on X having $\mathcal{S}$ as a subbase and turning each $f_i \colon (X, \mathcal{T}) \to (Y_i, \mathcal{T}_i)$ into a continuous function by Proposition 3.2.4.

Let $\mathcal{T}'$ be any topology on X such that $f_i \colon (X, \mathcal{T}') \to (Y_i, \mathcal{T}_i)$ is continuous for each $i \in \mathbb{I}$. By Proposition 3.2.4, it is clear that $\mathcal{S} \subset \mathcal{T}'$ and thus $\mathcal{T} \subset \mathcal{T}'$; that is. $\mathcal{T}$ is coarser than $\mathcal{T}'$. $\square$

The relevance of Proposition 3.2.7 becomes clear in the next section.

As we have seen at the beginning of this section, sequences are too limited an instrument in the study of topological spaces (and the continuous functions on them). Nevertheless, arguments involving sequences were often very convenient when we studied metric spaces in the previous chapter. Isn't there somehow a way to "rescue" sequences for the use in general topological spaces? There is, but it comes at a price: instead of just $\mathbb{N}$, we have to admit more general index sets.

Definition 3.2.8. *An ordered set $\mathbb{A}$ is called* directed *if, for any $\alpha, \beta \in \mathbb{A}$, there is $\gamma \in \mathbb{A}$ such that $\alpha \preceq \gamma$ and $\beta \preceq \gamma$.*

Examples 3.2.9. (a) Every totally ordered set is directed. In particular, $\mathbb{N}$ is directed.

(b) Let S be any set. Then $\mathfrak{P}(S)$, ordered by set inclusion, is directed.

Definition 3.2.10. *A* net *or a* generalized sequence *in a set S is a function from a directed set into S.*

If $\mathbb{A}$ is a directed set serving as the domain of some net, we often use the notation $(x_\alpha)_{\alpha \in \mathbb{A}}$; if no ambiguity can arise about $\mathbb{A}$, we sometimes simply write $(x_\alpha)_\alpha$. It is clear that a sequence is just a particular case of a net.

Example 3.2.11. Let $a < b$ be real numbers. A *partition* $\mathcal{P}$ of $[a, b]$ consists of finitely many numbers $t_0, t_1, \ldots, t_n$ such that $a = t_0 < t_1 < \cdots < t_n = b$. We write

$$\mathcal{P} = \{a = t_0 < t_1 < \cdots < t_n = b\}. \qquad (*)$$

The collection $\mathbb{P}$ of all partitions of $[a, b]$ is naturally ordered; given $\mathcal{P}$ as in $(*)$ and

$$\mathcal{Q} = \{a = s_0 < s_1 < \cdots < s_m = b\},$$

we define

$$\mathcal{P} \preceq \mathcal{Q} \quad :\Longleftrightarrow \quad \{t_0, t_1, \ldots, t_n\} \subset \{s_0, s_1, \ldots, s_m\}.$$

Clearly, this turns $\mathbb{P}$ into a directed set. For $\mathcal{P} \in \mathbb{P}$ as in $(*)$, a *tag* associated with $\mathcal{P}$ is an n-tuple $\xi = (\xi_1, \ldots, \xi_n)$ with $\xi_j \in [t_{j-1}, t_j]$ for $j = 1, \ldots, n$. Given $\mathcal{P} \in \mathbb{P}$, an associated tag ξ, and a function $f \colon [a, b] \to \mathbb{R}$, the corresponding *Riemann sum* is defined as

$$R(f; \mathcal{P}, \xi) := \sum_{j=1}^{n} f(\xi_j)(t_j - t_{j-1}).$$

If, for each $\mathcal{P} \in \mathbb{P}$, we fix a tag $\xi_\mathcal{P}$, the Riemann sums $(R(f; \mathcal{P}, \xi_\mathcal{P}))_{\mathcal{P} \in \mathbb{P}}$ form a net in $\mathbb{R}$.

It is clear how Definition 3.2.1 has to be extended to general nets.

Definition 3.2.12. *Let $(X, \mathcal{T})$ be a topological space. A net $(x_\alpha)_{\alpha \in \mathbb{A}}$ in X is said to* converge *to $x \in X$ if, for each $N \in \mathcal{N}_x$, there is $\alpha_N \in \mathbb{A}$ such that $x_\alpha \in N$ for all $\alpha \in \mathbb{A}$ such that $\alpha_N \preceq \alpha$. We then say that x is a* limit *of $(x_\alpha)_{\alpha \in \mathbb{A}}$ and write $x = \lim_\alpha x_\alpha$ or $x_\alpha \to x$.*

The notation $\lim_\alpha x_\alpha = x$ has to be handled with caution: a limit of a net in a general topological space need not be unique. An easy, albeit extreme, example is a chaotic topological space with more than one point: every net converges to every point. Writing $\lim_\alpha x_\alpha = x$ therefore does *not* mean that x is *the* limit of the net $(x_\alpha)_\alpha$, but rather that x is one (of possibly many) limits of that net.

We now give a few examples of convergent nets that may not be all that unfamiliar.

Examples 3.2.13. (a) In the situation of Example 3.2.11, suppose that f is Riemann integrable on $[a, b]$. Then $(R(f; \mathcal{P}, \xi_\mathcal{P}))_{\mathcal{P} \in \mathbb{P}}$ converges, namely

$$\lim_{\mathcal{P}} R(f; \mathcal{P}, \xi_\mathcal{P}) = \int_a^b f(t) \, dt.$$

Conversely, if $(R(f; \mathcal{P}, \xi_\mathcal{P}))_{\mathcal{P} \in \mathbb{P}}$ converges for each choice $(\xi_\mathcal{P})_{\mathcal{P} \in \mathbb{P}}$ of tags, and if this limit is independent of $(\xi_\mathcal{P})_{\mathcal{P} \in \mathbb{P}}$, then f is Riemann integrable.

(b) Let $S \neq \varnothing$ be a set, and let (Y, d) be a metric space. We say that a net $(f_\alpha)_{\alpha \in \mathbb{A}}$ in $F(S, Y)$ converges to $f \in F(S, Y)$ *pointwise* on S if $\lim_\alpha f_\alpha(x) = f(x)$, for each $x \in S$; that is, if for each $x \in S$ and for each $\epsilon > 0$, there is $\alpha_{x,\epsilon} \in \mathbb{A}$ with

$$d(f_\alpha(x), f(x)) < \epsilon \qquad (\alpha \in \mathbb{A}, \ \alpha_{x,\epsilon} \preceq \alpha);$$

and we say that $(f_\alpha)_{\alpha \in \mathbb{A}}$ converges to f *uniformly* on S if, for each $\epsilon > 0$, there is $\alpha_\epsilon \in \mathbb{A}$ with

$$d(f_\alpha(x), f(x)) < \epsilon \qquad (x \in S, \ \alpha \in \mathbb{A}, \ \alpha_\epsilon \preceq \alpha).$$

(In the situation where $S \subset \mathbb{R}^n$, $Y = \mathbb{R}$, and with $(f_\alpha)_{\alpha \in \mathbb{A}}$ a sequence, these definitions are just the familiar ones.)

We now show how to express these notions of convergence in terms of the topologies $\mathcal{T}_C$ (discussed in Example 3.1.11), where $\varnothing \neq C \subset \mathfrak{P}(X)$ is stable under finite unions. Let $(f_\alpha)_{\alpha \in \mathbb{A}}$ be a net in $F(S, Y)$, and let $f \in F(S, Y)$. Since the collection of sets $\{N_{f,C,\epsilon} : C \in \mathcal{C}, \ \epsilon > 0\}$ forms a base for $\mathcal{N}_f$, it is clear that $\lim_\alpha f_\alpha = f$ with respect to $\mathcal{T}_C$ if and only if, for each $C \in \mathcal{C}$ and for each $\epsilon > 0$, there is $\alpha_{C,\epsilon} \in \mathbb{A}$ such that

$$\sup_{x \in C} d(f_\alpha(x), f(x)) < \epsilon \qquad (\alpha \in \mathbb{A}, \ \alpha_{C,\epsilon} \preceq \alpha).$$

Let $\mathcal{F}$ be the collection of all finite subsets of S, and suppose that $(f_\alpha)_{\alpha \in \mathbb{A}}$ in $F(S, Y)$ converges pointwise to $f \in F(S, Y)$. Let $F \in \mathcal{F}$, say $F = \{x_1, \ldots, x_n\}$, and let $\epsilon > 0$. Then, for $j = 1, \ldots, n$, there is $\alpha_{j,\epsilon} \in \mathbb{A}$ such that

$$d(f_\alpha(x_j), f(x_j)) < \epsilon \qquad (\alpha \in \mathbb{A}, \ \alpha_{j,\epsilon} \preceq \alpha).$$

Since $\mathbb{A}$ is directed, there is $\alpha_{F,\epsilon} \in \mathbb{A}$ with $\alpha_{j,\epsilon} \preceq \alpha_{F,\epsilon}$ for $j = 1, \ldots, n$. It follows that

$$\max_{j=1,\ldots,n} d(f_\alpha(x_j), f(x_j)) < \epsilon \qquad (\alpha \in \mathbb{A}, \ \alpha_{F,\epsilon} \preceq \alpha),$$

so that we have convergence with respect to $\mathcal{T}_\mathcal{F}$. Conversely (and straightforward to verify), we have pointwise convergence whenever we have convergence with respect to $\mathcal{T}_\mathcal{F}$. We therefore sometimes call $\mathcal{T}_\mathcal{F}$ *the topology of pointwise convergence* (on S).

Similarly, uniform convergence is nothing but convergence with respect to $\mathcal{T}_{\{S\}}$, so that we sometimes refer to this topology as to *the topology of uniform convergence*.

As we saw at the beginning of this section, sequences are an inadequate instrument in the study of general topological spaces because, among other things, Proposition 2.3.4 is no longer true. As it turns out, this changes if we only replace sequences by nets.

Proposition 3.2.14. *Let $(X, \mathcal{T})$ be a topological space, and let $S \subset X$. Then $\overline{S}$ consists of those points in X that are a limit of a net in S.*

Proof. Let $x \in \overline{S}$. Turn $\mathcal{N}_x$ into a directed set via reversed set inclusion; that is, by letting

$$M \preceq N \quad :\Longleftrightarrow \quad N \subset M \qquad (N, M \in \mathcal{N}_x).$$

By Proposition 3.1.18, there is, for each $N \in \mathcal{N}_x$, an element $x_N \in N \cap S$. Then $(x_N)_{N \in \mathcal{N}_x}$ is a net in S such that $x = \lim_N x_N$.

Let $(x_\alpha)_{\alpha \in \mathbb{A}}$ be a net in S such that $x = \lim_\alpha x_\alpha$, and assume that $x \in U := X \setminus \overline{S}$. Then there is $\alpha_U \in \mathbb{A}$ such that $x_\alpha \in U \subset X \setminus S$ for all $\alpha \in \mathbb{A}$ such that $\alpha_U \preceq \alpha$, which is impossible. $\quad\square$

Corollary 3.2.15. *Let $(X, \mathcal{T})$ be a topological space. Then $F \subset X$ is closed if and only if every net in F that converges in X has its limits in F.*

Proposition 3.2.14 yields further examples for the nonuniqueness of limits in general topological spaces (and more natural ones than chaotic spaces).

Example 3.2.16. Let R be a commutative ring with identity that is an integral domain. The singleton subset $\{(0)\}$ is then dense in all of $\mathrm{Spec}(R)$ (Example 3.1.22). By Proposition 3.2.14, the constant net $((0))_\alpha$, no matter what the index set is, therefore converges to every point in $\mathrm{Spec}(R)$. If R is not a field (e.g., $R = \mathbb{Z}$), then $\mathrm{Spec}(R)$ has other elements besides (0). In this case, the net $((0))_\alpha$ has several limits, namely each point in $\mathrm{Spec}(R)$.

Our next proposition shows that Definition 3.1.3 is crucial when it comes to uniqueness of the limit in topological spaces.

Proposition 3.2.17. *The following are equivalent for a topological space $(X, \mathcal{T})$.*

(i) *X is Hausdorff.*
(ii) *Every convergent net in X has a unique limit.*

Proof. (i) $\Longrightarrow$ (ii): Let $(x_\alpha)_{\alpha \in \mathbb{A}}$ be a net in X such that there are $x, x' \in X$ such that $x_\alpha \to x$ and $x_\alpha \to x'$, but $x \neq x'$. By Definition 3.1.3, there are $N \in \mathcal{N}_x$ and $M \in \mathcal{N}_{x'}$ such that $N \cap M = \varnothing$. By the definition of convergence, there are $\alpha_N, \alpha_M \in \mathbb{A}$ such that $x_\alpha \in N$ for all $\alpha \in \mathbb{A}$ with $\alpha_N \preceq \alpha$ and $x_\alpha \in M$ for all $\alpha \in \mathbb{A}$ with $\alpha_M \preceq \alpha$. Since $\mathbb{A}$ is directed, we can find $\alpha \in \mathbb{A}$ with both $\alpha_N \preceq \alpha$ and $\alpha_M \preceq \alpha$. Consequently, $x_\alpha \in N \cap M$ must hold for such α, which is impossible because $N \cap M = \varnothing$.

(ii) $\Longrightarrow$ (i): Assume that X is not Hausdorff. Then there are $x, y \in X$ with $x \neq y$ such that $N \cap M \neq \varnothing$ for all $N \in \mathcal{N}_x$ and $M \in \mathcal{N}_y$. Turn $\mathcal{N}_x \times \mathcal{N}_y$ into a directed set by letting, for $(N_1, M_1), (N_2, M_2) \in \mathcal{N}_x \times \mathcal{N}_y$,

$$(N_1, M_1) \preceq (N_2, M_2) \quad :\Longleftrightarrow \quad N_2 \subset N_1 \text{ and } M_2 \subset M_1.$$

For any $(N, M) \in \mathcal{N}_x \times \mathcal{N}_y$ pick $x_{(N,M)} \in N \cap M$. It is routinely checked that $\big(x_{(N,M)}\big)_{(N,M) \in \mathcal{N}_x \times \mathcal{N}_y}$ is a net in X that converges to both x and y. $\quad\square$

As it turns out, with nets instead of sequences, we have an analogue of Theorem 2.3.7.

Theorem 3.2.18. *Let $(X, \mathcal{T}_X)$ and $(Y, \mathcal{T}_Y)$ be topological spaces, and let $x_0 \in X$. Then the following are equivalent for a function $f \colon X \to Y$.*

(i) *f is continuous at x_0.*
(ii) *For each net $(x_\alpha)_\alpha$ in X with $\lim_\alpha x_\alpha = x_0$, we have $\lim_\alpha f(x_\alpha) = f(x_0)$.*

Proof. (i) $\implies$ (ii): Let $(x_\alpha)_\alpha$ be a net in X with $\lim_\alpha x_\alpha = x_0$, and let $N \in \mathcal{N}_{f(x_0)}$. By the definition of continuity at a point, we have $f^{-1}(N) \in \mathcal{N}_{x_0}$. Consequently, there is an index $\alpha_{f^{-1}(N)}$ such that $x_\alpha \in f^{-1}(N)$ for all α such that $\alpha_{f^{-1}(N)} \preceq \alpha$. But this means that $f(x_\alpha) \in N$ for those α. Since $N \in \mathcal{N}_{f(x_0)}$ was arbitrary, this means that $\lim_\alpha f(x_\alpha) = f(x_0)$.

(ii) $\implies$ (i): Let $N \in \mathcal{N}_{f(x_0)}$, and assume towards a contradiction that $f^{-1}(N) \notin \mathcal{N}_{x_0}$. It follows that $U \not\subset f^{-1}(N)$ for each open subset U of X containing x_0. Let $\mathcal{U}_{x_0}$ denote the collection of all open subsets of X containing x_0. Then $\mathcal{U}_{x_0}$ is a directed subset of $\mathcal{N}_{x_0}$ (with respect to reversed set inclusion). By our assumption, we can choose $x_U \in U \setminus f^{-1}(N)$ for each $U \in \mathcal{U}_{x_0}$. It is clear that $\lim_U x_U = x_0$. However, since $f(x_U) \notin N$ for all $U \in \mathcal{U}_{x_0}$, it follows that $f(x_U) \not\to f(x_0)$. $\square$

We can use Theorem 3.2.18 to give an alternative description of the topology introduced in Proposition 3.2.7.

Proposition 3.2.19. *Let X be a set, let $((Y_i, \mathcal{T}_i))_{i \in \mathbb{I}}$ be a family of topological spaces, let $f_i \colon X \to Y_i$ be a function for each $i \in \mathbb{I}$, and let $\mathcal{T}$ be the coarsest topology on X such that each of the maps f_i is continuous. Then a net $(x_\alpha)_{\alpha \in \mathbb{A}}$ converges in $(X, \mathcal{T})$ to $x \in X$ if and only if $(f_i(x_\alpha))_{\alpha \in \mathbb{A}}$ converges to $f_i(x)$ in $(Y_i, \mathcal{T}_i)$ for each $i \in \mathbb{I}$.*

Proof. Suppose that $x = \lim_\alpha x_\alpha$ in $(X, \mathcal{T})$. Since each $f_i \colon X \to Y_i$ is continuous, it follows that $\lim_\alpha f_i(x_\alpha) = f_i(x)$ in $(Y_i, \mathcal{T}_i)$ for each $i \in \mathbb{I}$.

Conversely, suppose that $\lim_\alpha f_i(x_\alpha) = f_i(x)$ in $(Y_i, \mathcal{T}_i)$ for each $i \in \mathbb{I}$. Let N be a neighborhood of x in $(X, \mathcal{T})$. By the definition of a neighborhood, there is an open subset U of $(X, \mathcal{T})$ with $x \in U \subset N$. By the description of a subbase for $\mathcal{T}$ given in Proposition 3.2.7, there are $i_1, \ldots, i_n \in \mathbb{I}$ along with sets $U_j \in \mathcal{T}_{i_j}$ for $j = 1, \ldots, n$ such that

$$x \in f_{i_1}^{-1}(U_1) \cap \cdots \cap f_{i_n}^{-1}(U_n) \subset U.$$

Since $\lim_\alpha f_{i_j}(x_\alpha) = f_{i_j}(x)$ for $j = 1, \ldots, n$, there are $\alpha_1, \ldots, \alpha_n \in \mathbb{A}$ such that

$$f_{i_j}(x_\alpha) \in U_j \qquad (j = 1, \ldots, n, \ \alpha \in \mathbb{A}, \ \alpha_j \preceq \alpha).$$

Since $\mathbb{A}$ is directed, there is $\alpha_N \in \mathbb{A}$ such that $\alpha_j \preceq \alpha_N$ for $j = 1, \ldots, n$. Consequently, we have

$$x_\alpha \in f_{i_1}^{-1}(U_1) \cap \cdots \cap f_{i_n}^{-1}(U_n) \subset U \subset N \qquad (\alpha \in \mathbb{A}, \ \alpha_N \preceq \alpha),$$

so that $x = \lim_\alpha x_\alpha$ holds in $(X, \mathcal{T})$. $\square$

Exercises

1. Let $(X, \mathcal{T}_X)$, $(Y, \mathcal{T}_Y)$, and $(Z, \mathcal{T}_Z)$ be topological spaces, let $g : X \to Y$ be continuous at $x_0 \in X$, and let $f : Y \to Z$ be continuous at $g(x_0) \in Y$. Show that $f \circ g$ is continuous at x_0.
2. Let $(X, \mathcal{T}_X)$ and $(Y, \mathcal{T}_Y)$ be topological spaces, and let $\mathcal{B}_Y$ and $\mathcal{S}_Y$ be a base and a subbase, respectively, for $\mathcal{T}_Y$. Show that $f : X \to Y$ is continuous if and only if $f^{-1}(B) \in \mathcal{T}_X$ for each $B \in \mathcal{B}_Y$ and if and only if $f^{-1}(S) \in \mathcal{T}_X$ for each $S \in \mathcal{S}_Y$.
3. Let $(X, \mathcal{T}_X)$ and $(Y, \mathcal{T}_Y)$ be topological spaces, and let X_1 and X_2 be two subspaces of X such that $X_1 \cup X_2 = X$ which are both open or both closed. Show that $f : X \to Y$ is continuous if and only if $f|_{X_j}$ is continuous for $j = 1, 2$.
4. Let $(X, \mathcal{T})$ be a first countable topological space. Show that, for any $S \subset X$, the closure $\overline{S}$ of S consists of those points $x \in X$ such that there is a *sequence* in S converging to x.
5. Let R and S be commutative rings with identity, and let $\phi : R \to S$ be a unital ring homomorphism, that is, an additive and multiplicative map that maps the identity of R to the identity of S. Show that:
   (a) For any ideal I of S, its inverse image $\phi^{-1}(I)$ is an ideal of R; if I is proper or prime, respectively, the same is true for $\phi^{-1}(I)$;
   (b) The map
   $$\phi^* : \operatorname{Spec}(S) \to \operatorname{Spec}(R), \quad \mathfrak{p} \mapsto \phi^{-1}(\mathfrak{p})$$
   is continuous if both $\operatorname{Spec}(S)$ and $\operatorname{Spec}(R)$ are equipped with their respective Zariski topology.
6. Let $(X, \mathcal{T})$ be a topological space, and let $\approx$ be an equivalence relation on X. Show that the quotient topology on $X/\approx$ (see Exercise 3.1.8) is the finest topology on $X/\approx$ making the *quotient map*
   $$X \to X/\approx, \quad x \mapsto [x]$$
   continuous.
7. Let $(X, \mathcal{T})$ be a topological space, and let $f, g : X \to \mathbb{F}$ be continuous. Show that $f + g$, fg, and (provided that $f(x) \neq 0$ for $x \in X$) $\frac{1}{f}$ are also continuous.
8. Let $(X, \mathcal{T})$ be a topological space, and let $f, g : X \to \mathbb{R}$ be continuous. Show that $\max\{f, g\}$ and $\min\{f, g\}$ are continuous.
9. Let $(X, \mathcal{T}_X)$ and $(Y, \mathcal{T}_Y)$ be topological spaces, let $S \subset X$, and let $f : X \to Y$ be continuous. Show that $f\left(\overline{S}\right) \subset \overline{f(S)}$.
10. Let $(X, \mathcal{T}_X)$ and $(Y, \mathcal{T}_Y)$ be topological spaces such that Y is Hausdorff, let D be dense in X, and let $f, g : X \to Y$ be continuous functions such that $f|_D = g|_D$. Show that $f = g$. What happens if we drop the demand that Y be Hausdorff?
11. Let $\mathcal{F}$ denote the collection of all finite subsets of $[0, 1]$, and let $F([0, 1], \mathbb{F})$ be equipped with the topology $\mathcal{T}_{\mathcal{F}}$. Show that the continuous functions from $[0, 1]$ to $\mathbb{F}$ are dense in $(F([0, 1], \mathbb{F}), \mathcal{T}_{\mathcal{F}})$, but that there is no *sequence* of continuous functions from $[0, 1]$ to $\mathbb{F}$ converging to $f : [0, 1] \to \mathbb{F}$ given by
    $$f(t) := \begin{cases} 1, & t \notin [0, 1] \cap \mathbb{Q}, \\ 0, & t \in [0, 1] \cap \mathbb{Q}. \end{cases}$$

    (*Hint*: Recall that convergence in $\mathcal{T}_{\mathcal{F}}$ is pointwise convergence by Example 3.2.13(b), and use Exercise 2.4.7 to show that f cannot be the limit of a *sequence* of continuous functions.)

12. Let (X, d) be a metric space. A net $(x_\alpha)_{\alpha \in A}$ in X is called a *Cauchy net* if, for each $\epsilon > 0$, there is $\alpha_\epsilon \in A$ such that $d(x_\alpha, x_\beta) < \epsilon$ for all $\alpha, \beta \in A$ such that $\alpha_\epsilon \preceq \alpha, \beta$.
    (a) Show that every convergent net in X is a Cauchy net.
    (b) Show that X is complete if and only if each Cauchy net in X converges.

3.3 Compactness

Compactness for topological spaces is defined as in the metric situation.

Definition 3.3.1. *Let $(X, \mathcal{T})$ be a topological space, and let $S \subset X$. An* open cover *for S is a collection $\mathcal{U}$ of open subsets of X such that $S \subset \bigcup\{U : U \in \mathcal{U}\}$.*

Definition 3.3.2. *A subset K of a topological space $(X, \mathcal{T})$ is called* compact *if, for each open cover $\mathcal{U}$ of K, there are $U_1, \ldots, U_n \in \mathcal{U}$ such that $K \subset U_1 \cup \cdots \cup U_n$.*

Before we flesh out this definition with examples (nonmetrizable ones), we introduce yet another definition.

Definition 3.3.3. *A topological space $(X, \mathcal{T})$ has the* finite intersection property *if, for any collection $\mathcal{F}$ of closed subsets of X such that $\bigcap\{F : F \in \mathcal{F}\} = \varnothing$, there are $F_1, \ldots, F_n \in \mathcal{F}$ such that $F_1 \cap \cdots \cap F_n = \varnothing$.*

The following is straightforward (just pass to complements).

Proposition 3.3.4. *Let $(X, \mathcal{T})$ be a topological space. Then the following are equivalent.*

(i) *X is compact.*
(ii) *X has the finite intersection property.*

The reason why we introduced the finite intersection property at all is that it is sometimes easier to verify than compactness.

Example 3.3.5. Let R be a commutative ring with identity. We claim that $\mathrm{Spec}(R)$ has the finite intersection property (and thus is compact). Let $\mathcal{I}$ be a family of ideals of R such that

$$\bigcap\{V(I) : I \in \mathcal{I}\} = V\left(\sum\{I : I \in \mathcal{I}\}\right) = \varnothing.$$

This implies that $\sum\{I : I \in \mathcal{I}\} = R$: otherwise, Exercise 1.3.4 would yield a maximal ideal containing $\sum\{I : I \in \mathcal{I}\}$, and since maximal ideals are prime, this would be a contradiction. Since $1 \in \sum\{I : I \in \mathcal{I}\}$, there are $I_1, \ldots, I_n \in \mathcal{I}$ and $a_j \in I_j$ for $j = 1, \ldots, n$ such that $1 = \sum_{j=1}^n a_j$. It follows that $1 \in I_1 + \cdots + I_n$. Since the only ideal of R containing 1 is R itself, this means that $I_1 + \cdots + I_n = R$, so that

$$\varnothing = V(R) = V(I_1 + \cdots + I_n) = V(I_1) \cap \cdots \cap V(I_n).$$

Consequently, $\mathrm{Spec}(R)$ has the finite intersection property.

As can be expected, many of the properties of compact metric spaces do, in fact, hold for compact topological spaces.

Proposition 3.3.6. *Let $(X, \mathcal{T})$ be a topological space, and let Y be a subspace of X.*

(i) *If X is compact and Y is closed in X, then Y is compact.*
(ii) *If X is Hausdorff and Y is compact, then Y is closed in X.*

Proof. (i) is proved as for metric spaces (Proposition 2.5.4(i)).

For (ii), let $x \in X \setminus Y$. For each $y \in Y$, there are open subsets U_y, V_y of X such that $x \in U_y$, $y \in V_y$, and $U_y \cap V_y = \varnothing$. Since $\{V_y : y \in Y\}$ is an open cover for Y, there are $y_1, \ldots, y_n \in Y$ such that

$$Y \subset V_{y_1} \cup \cdots \cup V_{y_n}.$$

Letting $U := U_{y_1} \cap \cdots \cap U_{y_n}$, we obtain that

$$U \cap Y \subset U \cap (V_{y_1} \cup \cdots \cup V_{y_n}) = \varnothing,$$

so that $X \setminus Y$ is a neighborhood of x. Since $x \in X \setminus Y$ was arbitrary, this shows that $X \setminus Y$ is open. $\square$

The requirement in Proposition 3.3.6(ii) that X be Hausdorff cannot be dropped.

Example 3.3.7. Let R be a commutative ring with identity that is an integral domain, but not a field; for example, $R = \mathbb{Z}$. Then the singleton subset $\{(0)\}$ of $\mathrm{Spec}(R)$ is dense in R, but not closed (otherwise (0) would be the only prime ideal of R making it a field). Nevertheless, as a singleton subset, $\{(0)\}$ is trivially compact.

The following also holds (with an identical proof) as in the metric case.

Proposition 3.3.8. *Let $(K, \mathcal{T}_K)$ be a compact topological space, let $(Y, \mathcal{T}_Y)$ be any topological space, and let $f : K \to Y$ be continuous. Then $f(K)$ is compact.*

Corollary 3.3.9. *Let $(K, \mathcal{T}_K)$ be a nonempty, compact topological space, and let $f : K \to \mathbb{R}$ be continuous. Then f attains both a minimum and a maximum on K.*

We now use Proposition 3.3.8 to prove one of the most useful results on compact topological spaces. We first state another definition.

Definition 3.3.10. *Let $(X, \mathcal{T}_X)$ and $(Y, \mathcal{T}_Y)$ be topological spaces. A homeomorphism between X and Y is a bijective map $f : X \to Y$ such that both f and f^{-1} are continuous. If there is a homeomorphism between X and Y, the two spaces are called* homeomorphic.

To put it in a nutshell: two homeomorphic topological spaces cannot be told apart as topological spaces; that is., whatever topological property one space has is also enjoyed by the other.

Theorem 3.3.11. *Let $(K, \mathcal{T}_K)$ and $(Y, \mathcal{T}_Y)$ be topological spaces such that K is compact and Y is Hausdorff, and let $f: K \to Y$ be bijective and continuous. Then f is a homeomorphism.*

Proof. We verify Proposition 3.2.4(iii) for f^{-1}. Let $F \subset K$ be closed. Since f is bijective, it is clear that the inverse image of F under f^{-1} is just $f(F)$.

Since K is compact, so is F by Proposition 3.3.6(i). Consequently, $f(F)$ is compact in Y by Proposition 3.3.8 and thus closed in Y (because Y is Hausdorff) by Proposition 3.3.6(ii). This proves Proposition 3.2.4(iii) for f^{-1} and thus the continuity of f^{-1}. $\square$

Corollary 3.3.12. *Let X be a set, and let $\mathcal{T}_1$ and $\mathcal{T}_2$ be comparable topologies on X each turning it into a compact Hausdorff space. Then $\mathcal{T}_1 = \mathcal{T}_2$ holds.*

A metric space is compact if and only if each sequence in the space has a convergent subsequence. As we saw in the previous section, sequences are inadequate when it comes to dealing with general topological spaces. We now characterize the compact topological spaces in a similar way involving nets. We first need to define what we mean by a subnet.

Definition 3.3.13. *Let $\mathbb{A}$ and $\mathbb{B}$ be directed sets. A map $\phi: \mathbb{B} \to \mathbb{A}$ is called cofinal if, for each $\alpha \in \mathbb{A}$, there is $\beta_\alpha \in \mathbb{B}$ such that $\alpha \preceq \phi(\beta)$ for all $\beta \in \mathbb{B}$ such that $\beta_\alpha \preceq \beta$.*

Definition 3.3.14. *Let S be a nonempty set, and let $(x_\alpha)_{\alpha \in \mathbb{A}}$ and $(y_\beta)_{\beta \in \mathbb{B}}$ be nets in S. Then $(y_\beta)_{\beta \in \mathbb{B}}$ is a subnet of $(x_\alpha)_{\alpha \in \mathbb{A}}$ if $y_\beta = x_{\phi(\beta)}$ for a cofinal map $\phi: \mathbb{B} \to \mathbb{A}$.*

If $(y_\beta)_\beta$ is a subnet of $(x_\alpha)_\alpha$, we sometimes write $(x_{\alpha_\beta})_\beta$ instead of $(y_\beta)_\beta$ or $(x_{\phi(\beta)})_\beta$.

Of course, a subsequence of a sequence is a subnet. We want to stress, however, that a subnet of a sequence need not be a subsequence anymore: in Example 3.3.22 below, we encounter a sequence that has a convergent subnet, but no convergent subsequence.

Nevertheless, as for sequences, we have the following.

Proposition 3.3.15. *Let $(X, \mathcal{T})$ be a topological space, let $(x_\alpha)_{\alpha \in \mathbb{A}}$ be a net in X, and let $x \in X$ be a limit of $(x_\alpha)_{\alpha \in \mathbb{A}}$. Then each subnet of $(x_\alpha)_{\alpha \in \mathbb{A}}$ converges to x as well.*

Proof. Let $(y_\beta)_{\beta \in \mathbb{B}}$ be a subnet of $(x_\alpha)_\alpha$ with corresponding cofinal map $\phi: \mathbb{B} \to \mathbb{A}$.

Let $N \in \mathcal{N}_x$. Since $x = \lim_\alpha x_\alpha$, there is $\alpha_N \in \mathbb{A}$ such that $x_\alpha \in N$ for $\alpha \in \mathbb{A}$ with $\alpha_N \preceq \alpha$. Since ϕ is cofinal, there is $\beta_N \in \mathbb{B}$ such that $\alpha_N \preceq \phi(\beta)$ for all $\beta \in \mathbb{B}$ with $\beta_N \preceq \beta$. It follows that $y_\beta = x_{\phi(\beta)} \in N$ for all $\beta \in \mathbb{B}$ with $\beta_N \preceq \beta$. $\square$

The following definition and proposition are also helpful.

Definition 3.3.16. *Let* $(X, \mathcal{T})$ *be a topological space, and let* $(x_\alpha)_{\alpha \in \mathbb{A}}$ *be a net in* X. *A point* $x \in X$ *is an* accumulation point *for* $(x_\alpha)_{\alpha \in \mathbb{A}}$ *if, for each* $\alpha \in \mathbb{A}$ *and for each* $N \in \mathcal{N}_x$, *there is* $\beta \in \mathbb{A}$ *with* $\alpha \preceq \beta$ *such that* $x_\beta \in N$.

Proposition 3.3.17. *Let* $(X, \mathcal{T})$ *be a topological space, and let* $(x_\alpha)_{\alpha \in \mathbb{A}}$ *be a net in* X. *Then the following are equivalent for* $x \in X$.

(i) x *is an accumulation point of* $(x_\alpha)_{\alpha \in \mathbb{A}}$.
(ii) *There is a subnet of* $(x_\alpha)_{\alpha \in \mathbb{A}}$ *converging to* x.

Proof. (i) $\Longrightarrow$ (ii): Let $\mathbb{B} := \mathbb{A} \times \mathcal{N}_x$. For $(\alpha_1, N_1), (\alpha_2, N_2) \in \mathbb{B}$ define:

$$(\alpha_1, N_1) \preceq (\alpha_2, N_2) \quad :\Longleftrightarrow \quad \alpha_1 \preceq \alpha_2 \text{ and } N_2 \subset N_1.$$

This turns $\mathbb{B}$ into a directed set. Let $(\alpha, N) \in \mathbb{B}$. By the definition of an accumulation point, there is $\phi(\alpha, N) \in \mathbb{A}$ with $\alpha \preceq \phi(\alpha, N)$ such that $x_{\phi(\alpha, N)} \in N$. The map $\phi \colon \mathbb{B} \to \mathbb{A}$ is cofinal, and the net $\left(x_{\phi(\alpha, N)}\right)_{(\alpha, N) \in \mathbb{B}}$ converges to x.

(ii) $\Longrightarrow$ (i): Let $\mathbb{B}$ be a directed set, and let $\phi \colon \mathbb{B} \to \mathbb{A}$ be cofinal, such that $\left(x_{\phi(\beta)}\right)_{\beta \in \mathbb{B}}$ converges to x. Let $N \in \mathcal{N}_x$ and let $\alpha \in \mathbb{A}$. Since ϕ is cofinal, there is $\beta_\alpha \in \mathbb{B}$ such that $\alpha \preceq \phi(\beta)$ for all $\beta \in \mathbb{B}$ with $\beta_\alpha \preceq \beta$. Since $x = \lim_\beta x_{\phi(\beta)}$, there is $\beta_N \in \mathbb{B}$ such that $x_{\phi(\beta)} \in N$ for all $\beta \in \mathbb{B}$ with $\beta_N \preceq \beta$. Since $\mathbb{B}$ is directed, there is $\beta \in \mathbb{B}$ such that $\beta_\alpha \preceq \beta$ and $\beta_N \preceq \beta$. It follows that $\alpha \preceq \phi(\beta)$ and $x_{\phi(\beta)} \in N$. $\square$

We can now prove an analogue of Theorem 2.5.10 for general topological spaces.

Theorem 3.3.18. *For a topological space* $(X, \mathcal{T})$ *the following are equivalent:*

(i) X *is compact.*
(ii) *Each net in* X *has a convergent subnet.*

Proof. (i) $\Longrightarrow$ (ii): Let $(x_\alpha)_{\alpha \in \mathbb{A}}$ be a net in X. By Proposition 3.3.17, it is sufficient to show that $(x_\alpha)_{\alpha \in \mathbb{A}}$ has an accumulation point. Assume towards a contradiction that $(x_\alpha)_{\alpha \in \mathbb{A}}$ has no accumulation point. Then, for each $x \in X$, there is a neighborhood U_x of x (which we can choose to be open by making it smaller if necessary) and an index $\alpha_x \in \mathbb{A}$ such that $x_\alpha \notin U_x$ for all $\alpha \in \mathbb{A}$ such that $\alpha_x \preceq \alpha$. The collection $\{U_x : x \in X\}$ is an open cover of X. Since X is compact, there are $x_1, \ldots, x_n \in X$ such that

$$X = U_{x_1} \cup \cdots \cup U_{x_n}.$$

Choose $\alpha \in \mathbb{A}$ such that $\alpha_{x_j} \preceq \alpha$ for $j = 1, \ldots, n$, so that

$$x_\alpha \notin U_{x_1} \cup \cdots \cup U_{x_n} = X,$$

which is absurd.

(ii) $\implies$ (i): Assume that X is not compact. Then there is an open cover $\mathcal{U}$ of X that has no finite subcover. Let $\mathbb{U}$ be the collection of all finite subsets of $\mathcal{U}$ ordered by set inclusion. For each $v \in \mathbb{U}$, there is

$$x_v \in X \setminus \bigcup \{U : U \in v\} = \bigcap \{X \setminus U : U \in v\}$$

(otherwise, $\mathcal{U}$ would have a finite subcover). By hypothesis, the net $(x_v)_{v \in \mathbb{U}}$ has an accumulation point $x \in X$.

Fix $U \in \mathcal{U}$. By definition, $\{U\} \in \mathbb{U}$ holds. Therefore, by the definition of an accumulation point, there is, for any neighborhood N of x, an element $v_U \in \mathbb{U}$ with $\{U\} \preceq v_U$ (meaning $U \in v_U$) such that $x_{v_U} \in N$. Since also

$$x_{v_U} \in \bigcap \{X \setminus V : V \in v_U\} \subset X \setminus U,$$

it follows that $N \cap (X \setminus U) \neq \varnothing$. Assume that $x \in U$. Since U is then an open neighborhood of x, the preceding argument implies that $U \cap (X \setminus U) \neq \varnothing$, which is absurd. It follows that $x \in X \setminus U$.

Since $U \in \mathcal{U}$ is arbitrary, we eventually obtain

$$x \in \bigcap \{X \setminus U : U \in \mathcal{U}\} = X \setminus \bigcup \{U : U \in \mathcal{U}\} = \varnothing,$$

which is again absurd. $\square$

We shall see in Example 3.3.22 below that the corresponding statement about convergent sub*sequences* of *sequences* in compact spaces becomes false in the general topological setting.

We now prepare the ground for Tychonoff's theorem, one of the most fundamental results in set-theoretic topology.

Definition 3.3.19. *Let $((X_i, \mathcal{T}_i))_{i \in \mathbb{I}}$ be a family of topological spaces, and let $X := \prod_{i \in \mathbb{I}} X_i$. Then the* product topology *on X is the coarsest topology $\mathcal{T}$ on X making the* coordinate projections

$$\pi_i : X \to X_i, \quad f \mapsto f(i) \qquad (i \in \mathbb{I})$$

continuous. We call $(X, \mathcal{T})$ the topological product of $((X_i, \mathcal{T}_i))_{i \in \mathbb{I}}$.

By Proposition 3.2.7, the product topology does exist, and its open sets are the unions of sets of the form

$$\pi_{i_1}^{-1}(U_1) \cap \cdots \cap \pi_{i_n}^{-1}(U_n),$$

where $i_1, \ldots, i_n \in \mathbb{I}$ and $U_j \in \mathcal{T}_{i_j}$ for $j = 1, \ldots, n$. From Proposition 3.2.19 we know that a net $(f_\alpha)_\alpha$ converges to f in $(X, \mathcal{T})$ if and only if $(f_\alpha(i))_\alpha = (\pi_i(f_\alpha))_\alpha$ converges to $f(i) = \pi_i(f)$ in $(X_i, \mathcal{T}_i)$ for each $i \in \mathbb{I}$: the product topology is the *topology of coordinatewise convergence*. This second fact shows that, implicitly, we have already encountered the product topology in two special instances.

Examples 3.3.20. (a) Let $S \neq \varnothing$ be a set, and let (Y, d) be a metric space. Then $F(S, Y)$ is just another symbol for Y^S, and the product topology on Y^S is the topology of pointwise convergence, that is, $\mathcal{T}_{\mathcal{F}}$, where $\mathcal{F}$ is the collection of finite subsets of S.

(b) Let $((X_n, d_n))_{n=1}^{\infty}$ be a sequence of metric spaces. Then the product topology on $X := \prod_{n=1}^{\infty} X_n$ is metrizable via the metric

$$d((x_1, x_2, \ldots), (y_1, y_2, \ldots)) = \sum_{n=1}^{\infty} \frac{1}{2^n} \frac{d_n(x_n, y_n)}{1 + d(x_n, y_n)}$$

for $(x_1, x_2, \ldots), (y_1, y_2, \ldots) \in X$; this follows from Exercise 2.3.1.

Theorem 3.3.21 (Tychonoff's theorem). *Let $((K_i, \mathcal{T}_i))_{i \in \mathbb{I}}$ be a nonempty family of compact topological spaces. Then their topological product is also compact.*

Proof. Let $(f_\alpha)_\alpha$ be a net in $K := \prod_{i \in \mathbb{I}} K_i$. Let $\mathbb{J} \subset \mathbb{I}$ be nonempty and let $f \in K$. We call $(\mathbb{J}, f)$ a *partial accumulation point* of $(f_\alpha)_\alpha$ if $f|_{\mathbb{J}}$ is an accumulation point of $(f_\alpha|_{\mathbb{J}})_\alpha$ in $\prod_{j \in \mathbb{J}} K_j$. Obviously, a partial accumulation point $(\mathbb{J}, f)$ is an accumulation point of $(f_\alpha)_\alpha$ if and only if $\mathbb{J}$ is all of $\mathbb{I}$.

Let $\mathcal{P}$ be the set of all partial accumulation points of $(f_\alpha)_\alpha$. For any two $(\mathbb{J}_f, f), (\mathbb{J}_g, g) \in \mathcal{P}$, define

$$(\mathbb{J}_f, f) \preceq (\mathbb{J}_g, g) \quad :\Longleftrightarrow \quad \mathbb{J}_f \subset \mathbb{J}_g \text{ and } g|_{\mathbb{J}_f} = f.$$

Since K_i is compact for each $i \in \mathbb{I}$, the net $(f_\alpha)_\alpha$ has partial accumulation points $(\{i\}, f_i)$ for each $i \in \mathbb{I}$ by Theorem 3.3.18; in particular, $\mathcal{P}$ is not empty.

Let $\mathcal{Q}$ be a totally ordered subset of $\mathcal{P}$. Let $\mathbb{J}_g := \bigcup \{\mathbb{J}_f : (\mathbb{J}_f, f) \in \mathcal{Q}\}$. Define $g \in K$ by letting $g(j) := f(j)$ for $j \in \mathbb{J}_f$ with $(\mathbb{J}_f, f) \in \mathcal{Q}$ (and arbitrarily on $\mathbb{I} \setminus \mathbb{J}_g$). Since $\mathcal{Q}$ is totally ordered, g is well defined. We claim that $(\mathbb{J}_g, g)$ is a partial accumulation point of $(f_\alpha)_\alpha$. Let $N \subset \prod_{j \in \mathbb{J}_g} K_j$ be a neighborhood of $g|_{\mathbb{J}_g}$. By Proposition 3.2.7, we may suppose that

$$N = \pi_{j_1}^{-1}(U_{j_1}) \cap \cdots \cap \pi_{j_n}^{-1}(U_{j_n}),$$

where $j_1, \ldots, j_n \in \mathbb{J}_g$, and $U_{j_1} \subset K_{j_1}, \ldots, U_{j_n} \subset K_{j_n}$ are open. Let $(\mathbb{J}_h, h) \in \mathcal{Q}$ be such that $\{j_1, \ldots, j_n\} \subset \mathbb{J}_h$ (this is possible because $\mathcal{Q}$ is totally ordered). Since $(\mathbb{J}_h, h)$ is a partial accumulation point of $(f_\alpha)_\alpha$, it follows that there is, for each index α, an index β with $\alpha \preceq \beta$ and

$$f_\beta(j_k) = \pi_{j_k}(f_\beta) \in U_{j_k} \qquad (k = 1, \ldots, n),$$

so that $f_\beta \in N$. Hence, $(\mathbb{J}_g, g)$ is indeed a partial accumulation point of $(f_\alpha)_\alpha$ and thus lies in $\mathcal{P}$.

By Zorn's lemma, $\mathcal{P}$ has a maximal element $(\mathbb{J}_{\max}, f_{\max})$. Assume that $\mathbb{J}_{\max} \subsetneq \mathbb{I}$; that is, there is $i_0 \in \mathbb{I} \setminus \mathbb{J}_{\max}$. Since $(\mathbb{J}_{\max}, f_{\max})$ is a partial accumulation point of $(f_\alpha)_\alpha$, there is a subnet $(f_{\alpha_\beta})_\beta$ of $(f_\alpha)_\alpha$ such that

$\pi_j(f_{\alpha_\beta}) \to \pi_j(f_{\max})$ for each $j \in \mathbb{J}_{\max}$. Since K_{i_0} is compact, we may find a subnet $\left(f_{\alpha_{\beta_\gamma}}\right)_\gamma$ of $(f_{\alpha_\beta})_\beta$ such that $\pi_{i_0}(f_{\alpha_{\beta_\gamma}})_\gamma$ converges to some x_{i_0} in K_{i_0}. Define $\tilde{f} \in K$ by letting $\tilde{f}|_{\mathbb{J}_{\max}} = f_{\max}$ and $\tilde{f}(i_0) = x_{i_0}$. It follows that $\left(\mathbb{J}_{\max} \cup \{i_0\}, \tilde{f}\right)$ is a partial accumulation point of $(f_\alpha)_\alpha$, which contradicts the maximality of $(\mathbb{J}_{\max}, f_{\max})$. $\square$

We now use Tychonoff's theorem to exhibit a compact topological space and a sequence in that space without a convergent subsequence:

Example 3.3.22. Let $\mathbb{I}$ be the set of all strictly increasing sequences in $\mathbb{N}$, that is, sequences $(n_k)_{k=1}^\infty$ such that $n_1 < n_2 < n_3 < \cdots$. It is not difficult to see that $\mathbb{I}$ has cardinality $\mathfrak{c}$; we won't need this, however. For each $i = (n_k)_{k=1}^\infty$ in $\mathbb{I}$, define a sequence $(f_n(i))_{n=1}^\infty$ as follows,

$$f_n(i) := \begin{cases} (-1)^k, & \text{if } n = n_k \text{ for some } k \in \mathbb{N}, \\ 0, & \text{otherwise.} \end{cases}$$

Then $(f_n(i))_{n=1}^\infty$ is a sequence in $[-1, 1]$ such that the subsequence $(f_{n_k}(i))_{k=1}^\infty$ diverges. The topological product $[-1, 1]^{\mathbb{I}}$ is compact by Tychonoff's theorem. Assume that the sequence $(f_n)_{n=1}^\infty$ has a convergent subsequence, say $(f_{n_k})_{k=1}^\infty$. By the definition of the product topology, this means that $(f_{n_k}(i))_{k=1}^\infty$ converges for each $i \in \mathbb{I}$. For $i = (n_k)_{k=1}^\infty$, this is impossible according to our construction. Of course, the sequence $(f_n)_{n=1}^\infty$ has a convergent sub*net*.

Another application of Tychonoff's theorem, namely a proof of the Arzelà–Ascoli theorem, is given in Appendix C.

Concluding this section, we deal with a class of possibly noncompact topological spaces which, nevertheless, are reasonably close to being compact.

Definition 3.3.23. *Let $(X, \mathcal{T})$ be a topological space. Then X is said to be locally compact if $\mathcal{N}_x$ contains a compact subset of X for each $x \in X$.*

Examples 3.3.24. (a) Compact spaces are (obviously and trivially) locally compact.
(b) $\mathbb{R}^n$ is locally compact, but fails to be compact.
(c) Every discrete topological space is locally compact.
(d) Let E be the linear space $C([0, 1], \mathbb{F})$ equipped with the norm $\|\cdot\|_\infty$, and assume that it is locally compact. Then 0 has a compact neighborhood, say K. By the definition of a neighborhood, there is $\epsilon > 0$ such that $B_\epsilon(0) \subset K$, and since K is closed in E, it follows that $B_\epsilon[0] = \overline{B_\epsilon(0)} \subset K$ as well. Consequently, $B_\epsilon[0]$ must be compact. Since

$$B_\epsilon[0] \to B_1[0], \quad x \mapsto \frac{1}{\epsilon}x$$

is a homeomorphism, this implies that $B_1[0]$ is compact, too. As we have seen in Example 2.5.13, this is not true. (Invoking Theorem B.5 instead of Example 2.5.13, one can see that no infinite-dimensional, normed space is locally compact.)

Nevertheless, even though $\mathbb{R}^n$ is not compact, it is not far away from a certain compact space. We illustrate this in the case where $n = 2$.

Example 3.3.25. We may identify $\mathbb{C} \cong \mathbb{R}^2$ with the xy-plane in $\mathbb{R}^3$. Let $\mathbb{C}_\infty$ be the unit sphere in $\mathbb{R}^3$ (i.e., those vectors with Euclidean length one): this space is compact. Let $p = (0,0,1)$. Then, for each $z = x + iy \in \mathbb{C}$, there is a unique line in $\mathbb{R}^3$ connecting $(x,y,0)$ with p. Let $\iota(z)$ be the intersection point of this line with $\mathbb{C}_\infty$. Then $\iota \colon \mathbb{C} \to \mathbb{C}_\infty$ is a continuous injective map whose range is $\mathbb{C}_\infty \setminus \{p\}$ and that is a homeomorphism onto its image. (All this needs to be checked, of course.) The sphere $\mathbb{C}_\infty$ is sometimes referred to as the *Riemann sphere*.

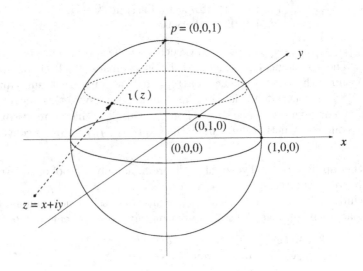

Fig. 3.2: Riemann sphere

Interestingly, a similar construction works for arbitrary locally compact Hausdorff spaces.

Theorem 3.3.26. *Let $(X,\mathcal{T})$ be a locally compact Hausdorff space. Then there is a compact Hausdorff space $(X_\infty, \mathcal{T}_\infty)$, the one-point compactification of X, along with a map $\iota \colon X \to X_\infty$ such that*

(i) *ι is a homeomorphism onto its image, and*
(ii) *$X_\infty \setminus \iota(X)$ consists of just one point.*

Moreover, $(X_\infty, \mathcal{T}_\infty)$ is unique up to homeomorphism.

Proof. Let ∞ be any point not contained in X, and let $X_\infty := X \cup \{\infty\}$. Define

$$\mathcal{T}_\infty := \mathcal{T} \cup \{X_\infty \setminus K : K \subset X \text{ is compact}\}.$$

From this definition, it is clear that $\mathcal{T} \subset \mathcal{T}_\infty$ and that $X \cap U \in \mathcal{T}$ for each $U \in \mathcal{T}_\infty$ (since X is Hausdorff, all its compact subsets are closed).

We claim that $\mathcal{T}_\infty$ is a topology on X_∞. It is clear that $\varnothing, X_\infty \in \mathcal{T}_\infty$.

Let $\mathcal{U}$ be an arbitrary family of sets in $\mathcal{T}_\infty$. If $\mathcal{U} \subset \mathcal{T}$, nothing has to be shown. Hence, we may suppose that there is a compact subset K of X such that $X_\infty \setminus K \in \mathcal{U}$. Since $K \cap U \in \mathcal{T}|_K$ for each $U \in \mathcal{U}$, there is $V \in \mathcal{T}$ such that

$$\bigcup \{K \cap U : U \in \mathcal{U}\} = K \cap V.$$

It follows that

$$\begin{aligned}
\bigcup \{U : U \in \mathcal{U}\} &= \bigcup \{K \cap U : U \in \mathcal{U}\} \cup (X_\infty \setminus K) \\
&= (K \cap V) \cup (X_\infty \setminus K) \\
&= V \cup (X_\infty \setminus K) \\
&= X_\infty ((X \setminus V) \cap K).
\end{aligned}$$

Since K is compact and $X \setminus V$ is closed, the subset $(X \setminus V) \cap K$ of X is closed in K and thus compact. It follows that $\bigcup \{U : U \in \mathcal{U}\} \in \mathcal{T}_\infty$.

Let $U_1, U_2 \in \mathcal{T}_\infty$. If $U_1, U_2 \subset X$, nothing has to be shown. Hence, suppose that $U_2 = X_\infty \setminus K_2$ for some compact subset K_2 of X. If $U_1 \subset X$, it follows that

$$U_1 \cap U_2 = U_1 \cap (X_\infty \setminus K_2) = U_1 \cap (X \setminus K_2),$$

which belongs to $\mathcal{T}$ because K_2 is closed. If $U_1 \not\subset X$, there is a compact set $K_1 \subset X$ with $U_1 = X_\infty \setminus K_1$. Since $K_1 \cup K_2$ is again compact (see Exercise 1 below), it follows that

$$U_1 \cap U_2 = X_\infty \setminus (K_1 \cup K_2) \in \mathcal{T}_\infty.$$

All in all, $\mathcal{T}_\infty$ is a topology on X_∞. By the remarks at the beginning of this proof, $\mathcal{T}_\infty|_X$ is just $\mathcal{T}$, so that the canonical embedding $\iota : X \to X_\infty$ is indeed a homeomorphism onto its image.

We claim that $(X_\infty, \mathcal{T}_\infty)$ is Hausdorff. To see this let $x, y \in X_\infty$ be such that $x \neq y$. Without loss of generality, suppose that $y = \infty$. Let $K \subset X$ be a compact neighborhood of x. Then there is $U \in \mathcal{T}$ with $x \in U \subset K$. Let $V := X_\infty \setminus K$. Then $V \in \mathcal{T}_\infty$ with $\infty \in V$, and $U \cap V = \varnothing$ holds.

Let $\mathcal{U}$ be an open cover for X_∞. Then there is $U_0 \in \mathcal{U}$ such that $\infty \in U_0$. Consequently, there is a compact subset K of X with $U_0 = X_\infty \setminus K$. Since $\{X \cap U : U \in \mathcal{U}\}$ is an open cover for K, there are $U_1, \ldots, U_n \in \mathcal{U}$ with $K \subset U_1 \cup \cdots \cup U_n$. It follows that

$$X_\infty = U_0 \cup K = U_0 \cup U_1 \cup \cdots \cup U_n.$$

Consequently, X_∞ is compact.

Let $(X'_\infty, \mathcal{T}'_\infty)$ be any other compact Hausdorff space containing a homeomorphic copy of X such that $X'_\infty \setminus X$ consists of only one element, say ∞'. Define $f : X_\infty \to X'_\infty$ as the identity on X and such that $f(\infty) := \infty'$, and let $U \in \mathcal{T}'_\infty$. Then there are two possibilities: $U \in \mathcal{T}$ or $\infty' \in U$. If $U \in \mathcal{T}$, then $f^{-1}(U) = U \in \mathcal{T} \subset \mathcal{T}_\infty$ holds. Suppose that $\infty' \in U$. Then $K := X_\infty \setminus U$ is a subset of X that is closed in X'_∞ and thus compact. It follows that $f^{-1}(U) = X_\infty \setminus K \in \mathcal{T}_\infty$. Hence, in either case, $f^{-1}(U)$ belongs to $\mathcal{T}_\infty$. Since $U \in \mathcal{T}'_\infty$ was arbitrary, this means that f is continuous. Reversing the rôles of X_∞ and X'_∞ yields the continuity of f^{-1}, so that f is a homeomorphism. This proves the uniqueness part of the theorem. $\quad\square$

Note that this construction also works when X is already compact.

Exercises

1. Let $(X, \mathcal{T})$ be a topological space, and let $K_1, \ldots, K_n$ be compact subsets of X. Show that $K_1 \cup \cdots \cup K_n$ is compact.

2. Let $(X, \mathcal{T})$ be a topological space, and let $\mathcal{B}$ be a base for $\mathcal{T}$. Show that X is compact if and only if each open cover $\mathcal{U} \subset \mathcal{B}$ of X has a finite subcover.

3. *Dini's lemma.* Let $(K, \mathcal{T})$ be a compact topological space, let $f : K \to \mathbb{R}$ be continuous, and let $(f_\alpha)_{\alpha \in \mathbb{A}}$ be a net of continuous functions from K to $\mathbb{R}$ such that $f(x) = \lim_\alpha f_\alpha(x)$ for all $x \in K$ and $f_\beta(x) \leq f_\alpha(x)$ for all $x \in K$ and $\alpha, \beta \in \mathbb{A}$ with $\alpha \preceq \beta$. Show that $(f_\alpha)_\alpha$ converges to f uniformly on K. (*Hint:* Fix $\epsilon > 0$ and consider, for each $\alpha \in \mathbb{A}$, the set $U_\alpha := \{x \in K : 0 \leq f_\alpha(x) - f(x) < \epsilon\}$.)

4. Let $(X, \mathcal{T}_X)$ be a topological space, let $((Y_i, \mathcal{T}_i))_{i \in \mathbb{I}}$ be a family of topological spaces, and let $(Y, \mathcal{T}_Y)$ denote its topological product. Show that $f : X \to Y$ is continuous if and only if $\pi_i \circ f : X \to Y_i$ is continuous for each $i \in \mathbb{I}$, where $\pi_i : Y \to Y_i$ is the coordinate projection.

5. Let $(X_1, \mathcal{T}_1), \ldots, (X_n, \mathcal{T}_n)$ be topological spaces, and, for $j = 1, \ldots, n$, let $\mathcal{B}_j$ be a base for $\mathcal{T}_j$. Show that the subsets of $X_1 \times \cdots \times X_n$ of the form $B_1 \times \cdots \times B_n$ with $B_j \in \mathcal{B}_j$ for $j = 1, \ldots, n$ form a base for the product topology on $X_1 \times \cdots \times X_n$.

6. Let $((X_i, \mathcal{T}_i))_{i \in \mathbb{I}}$ be a family of topological spaces. The sets of the form $\prod_{i \in \mathbb{I}} U_i$, where $U_i \in \mathcal{T}_i$ for $i \in \mathbb{I}$, form a base for a topology on $\prod_{i \in \mathbb{I}} X_i$, the *box topology*. Show that the box topology is finer than the product topology and that the two topologies coincide if and only if $(X_i, \mathcal{T}_i)$ is chaotic for all but finitely many $i \in \mathbb{I}$.

7. Let $((X_i, \mathcal{T}_i))_{i \in \mathbb{I}}$ be a family of topological spaces, and let $(X, \mathcal{T})$ be their topological product. Show that $(X, \mathcal{T})$ is Hausdorff if and only if $(X_i, \mathcal{T}_i)$ is Hausdorff for each $i \in \mathbb{I}$.

8. Let $(X, \mathcal{T})$ be a topological space, and let $X \times X$ be equipped with the product topology. Show that $\{(x, x) : x \in X\}$ is closed in $X \times X$ if and only if X is Hausdorff.

9. Let E be a normed space over $\mathbb{F} = \mathbb{R}$ or $\mathbb{F} = \mathbb{C}$. A *linear functional* on E is a map $\phi : E \to \mathbb{F}$ such that $\phi(\lambda x + \mu y) = \lambda\,\phi(x) + \mu\,\phi(y)$ for all $\lambda, \mu \in \mathbb{F}$ and $x, y \in E$. Let $\mathcal{F}$ denote the collection of all finite subsets of E, and let

$$K := \left\{ \phi \in \mathbb{F}^E : \phi \text{ is a linear functional on } E \text{ with } |\phi(x)| \leq \|x\| \text{ for } x \in E \right\}.$$

Prove the *Alaoglu–Bourbaki theorem*: $(K, \mathcal{T}_\mathcal{F}|_K)$ is compact.

10. Show that $\mathbb{Q}$ with the topology inherited from $\mathbb{R}$ is not locally compact.

3.4 Connectedness

Intuitively, one would call a topological space connected if one can "walk" from any point of the space to any other point of the space without having to "jump" over a "gap". How can this be made precise?

Definition 3.4.1. *Let $(X, \mathcal{T})$ be a topological space. A* path *in X is a continuous map from $[0,1]$ to X.*

Of course, any nondegenerate, closed, and bounded interval in $\mathbb{R}$ may serve as a domain for a path.

Definition 3.4.2. *Let $(X, \mathcal{T})$ be a topological space, and let $x_0, x_1 \in X$. Then x_0 and x_1 can be* connected by a path *in X if there is a path $\gamma \colon [0,1] \to X$ such that $\gamma(0) = x_0$ and $\gamma(1) = x_1$. We say that γ* connects *x_0 with x_1.*

Definition 3.4.3. *A topological space $(X, \mathcal{T})$ is called* path connected *if any two of its points can be connected by a path in X.*

Example 3.4.4. Let E be a normed space. Recall that a subset C of E is called *convex* if, for any $x, y \in C$, the line segment $\{x + t(y - x) : t \in [0,1]\}$ also lies in C. Trivially, the whole space is convex as are all its singleton subsets. The same is true for all open balls. Let $x_0 \in E$, let $r > 0$, and let $x, y \in B_r(x_0)$; then we have

$$\begin{aligned}
\|x + t(y - x) - x_0\| &= \|(1-t)x + ty - (1-t)x_0 - tx_0\| \\
&\leq (1-t)\|x - x_0\| + t\|y - x_0\| \\
&< r \qquad (t \in [0,1]).
\end{aligned}$$

Similarly, one sees that all closed balls are convex. Every convex subset C of E is path connected: for $x_0, x_1 \in C$, the function

$$\gamma \colon [0,1] \to E, \quad t \mapsto x_0 + t(x_1 - x_0)$$

is a path in C connecting x_0 with x_1. In particular, all intervals in $\mathbb{R}$—possibly unbounded or degenerate—are path connected.

Here comes an easily proven hereditary property for path connectedness.

Proposition 3.4.5. *Let $(X, \mathcal{T}_X)$ and $(Y, \mathcal{T}_Y)$ be topological spaces such that X is path connected, and let $f \colon X \to Y$ be continuous. Then $f(X)$ is path connected.*

Proof. Let $y_0, y_1 \in f(X)$, and let $x_0, x_1 \in X$ be such that $y_j = f(x_j)$ for $j = 0, 1$. Let $\gamma \colon [0,1] \to X$ be a path connecting x_0 with x_1. Then $f \circ \gamma$ is a path in $f(X)$ that connects y_0 with y_1. $\square$

Examples 3.4.6. (a) For $n \in \mathbb{N}$, let $\mathbb{S}^{n-1}$ denote the unit sphere in $\mathbb{R}^n$, that is, the collection of all vectors in $\mathbb{R}^n$ with Euclidean length one. We claim that $\mathbb{S}^{n-1}$ is path connected for $n \geq 2$, which we prove by induction on n. For $n = 2$, this is easy to see: the convex subset $[0, 2\pi]$ of $\mathbb{R}$ is path connected as is, by Proposition 3.4.5, $\mathbb{S}^1$ as the range of

$$[0, 2\pi] \to \mathbb{R}^2, \quad \theta \mapsto (\cos \theta, \sin \theta).$$

Suppose that we have already established the path connectedness of $\mathbb{S}^{n-1}$ for some $n \geq 2$. Then $\mathbb{S}^n$ is the range of the continuous map

$$[0, 2\pi] \times \mathbb{S}^{n-1} \to \mathbb{R}^{n+1}, \quad (\theta, x) \mapsto (\cos \theta, (\sin \theta)x).$$

Since $[0, 2\pi] \times \mathbb{S}^{n-1}$ is path connected by the induction hypothesis and Exercise 2 below, it follows from Proposition 3.4.5 that $\mathbb{S}^n$ is path connected.

(b) The subspace

$$\left\{ \left(x, \sin\left(\frac{1}{x}\right) \right) : x \in (0, 1] \right\}$$

of $\mathbb{R}^2$ is the image of the path connected space $(0, 1]$ under the continuous map

$$(0, 1] \to \mathbb{R}^2, \quad x \mapsto \left(x, \sin\left(\frac{1}{x}\right) \right)$$

and therefore path connected.

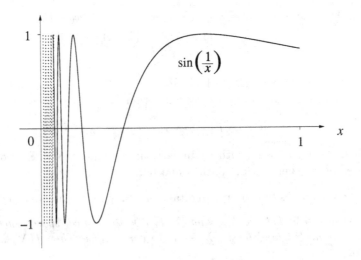

Fig. 3.3: Graph of $\sin\left(\frac{1}{x}\right)$ for $x \in (0, 1]$

Even though path connectedness is probably the most intuitive notion of connectedness it is often too restrictive to work with. Here comes the "proper" definition of connectedness.

Definition 3.4.7. *A topological space $(X, \mathcal{T})$ is said to be connected if there are no $U, V \in \mathcal{T}$—both nonempty—such that $U \cap V = \varnothing$ and $U \cup V = X$. Otherwise, X is called disconnected.*

If X is not connected, then there are nonempty open subsets U and V of X such that $U \cap V = \varnothing$ and $U \cup V = X$. It follows that both U and V are also closed. Subsets of topological spaces that are both open and closed are often referred to as *clopen*. Clearly, a topological space X is connected if and only if its only clopen subsets are the trivial ones: $\varnothing$ and X.

Examples 3.4.8. (a) We claim that $[0, 1]$ is connected. Otherwise, there are nonempty open sets $U, V \subset [0, 1]$ such that $U \cap V = \varnothing$ and $U \cup V = [0, 1]$. Without loss of generality, suppose that $0 \in U$, so that $0 \notin V$. Since U is open in $[0, 1]$, there is $\epsilon > 0$ such that $[0, \epsilon) \subset U$; it follows that $b := \inf V \geq \epsilon > 0$. Since V is also closed in $[0, 1]$, it is easy to see that $b \in V$. Let $a := \sup\{t \in U : t < b\}$. Clearly, $a \leq b$ holds, and since U is closed in $[0, 1]$, we have $a \in U$ and thus $a < b$. This, however, means that $(a, b) \subset [0, 1] \setminus (U \cup V)$, which is impossible.

(b) Any discrete space with more than one point is disconnected.

(c) Let $(X, \mathcal{T})$ be any topological space, and let $S, T \subset X$ be nonempty such that there are $U, V \in \mathcal{T}$ with $S \subset U$, $T \subset V$, and $U \cap V = \varnothing$. Then the subspace $S \cup T$ of X is disconnected (this subsumes the previous example). For instance, $(-\infty, -2] \cup [-1, 2) \cup [7, \pi^2]$ is a disconnected subspace of $\mathbb{R}$.

What does connectedness in the sense of Definition 3.4.7 have to do with path connectedness? Here is the answer.

Proposition 3.4.9. *Let $(X, \mathcal{T})$ be a path connected topological space. Then X is connected.*

Proof. Assume that X is not connected. Then there are nonempty open subsets U and V of X such that $U \cap V = \varnothing$ and $U \cup V = X$. Let $x_0 \in U$ and $x_1 \in V$. Since X is path connected, x_0 and x_1 are connected by a path, say γ. Then, however, $\gamma^{-1}(U)$ and $\gamma^{-1}(V)$ are open, nonempty, with an empty intersection, and their union is all of $[0, 1]$. Hence, $[0, 1]$ is not connected, which is impossible by the preceding example. $\square$

Example 3.4.10. Clearly, every interval in $\mathbb{R}$ (possibly unbounded or degenerate) is path connected and thus connected. Conversely, let $\varnothing \neq I \subset \mathbb{R}$ be any connected subspace. If I is not an interval, then there is $c \in \mathbb{R} \setminus I$ such that $(-\infty, c) \cap I \neq \varnothing$ and $(c, \infty) \cap I \neq \varnothing$. Clearly, $U := I \cap (-\infty, c)$ and $V := I \cap (c, \infty)$ are open in I such that $U \cap V = \varnothing$ and $U \cup V = I$. Therefore, the only connected (or, equivalently, path connected) subsets of $\mathbb{R}$ are the intervals.

There is an analogue of Proposition 3.4.5 for connectedness.

Proposition 3.4.11. *Let $(X, \mathcal{T}_X)$ and $(Y, \mathcal{T}_Y)$ be topological spaces such that X is connected, and let $f \colon X \to Y$ be continuous. Then $f(X)$ is connected.*

Proof. Let $U, V \in \mathcal{T}_Y|_{f(X)}$ be such that $U \cap V = \varnothing$ and $U \cup V = f(X)$. Then $f^{-1}(U)$ and $f^{-1}(V)$ are in $\mathcal{T}_X$ such that $f^{-1}(U) \cap f^{-1}(V) = \varnothing$ and $f^{-1}(U) \cup f^{-1}(V) = X$. Since X is connected, this is possible only if $f^{-1}(U)$ or $f^{-1}(V)$ is empty, which, in turn, is possible only if U or V is empty. Hence, $f(X)$ is connected. $\square$

Corollary 3.4.12 (intermediate value theorem). *Let $(X, \mathcal{T})$ be a connected topological space, and let $f \colon X \to \mathbb{R}$ be continuous. Then, for any $x_1, x_2 \in X$ and $c \in \mathbb{R}$ with $f(x_1) \leq c \leq f(x_2)$, there is $x_0 \in X$ with $f(x_0) = c$.*

Proof. Since $f(X)$ is a connected subspace of $\mathbb{R}$, it is an interval and thus contains all of $[f(x_1), f(x_2)]$. $\square$

Is path connectedness actually *really* stronger than mere connectedness? To answer this question, we prove another proposition.

Proposition 3.4.13. *Let $(X, \mathcal{T})$ be a topological space, and let Y be a dense connected subspace of X. Then X is connected.*

Proof. Let $U, V \in \mathcal{T}$ be such that $U \cap V = \varnothing$ and $U \cup V = X$. Since $Y \cap U$, $Y \cap V \in \mathcal{T}|_Y$, $(Y \cap U) \cap (Y \cap V) = \varnothing$, and $(Y \cap U) \cup (Y \cup V) = Y$, the connectedness of Y yields that $Y \cap U = Y$ or $Y \cap V = Y$; without loss of generality, suppose that $Y \cap U = Y$ (i.e., $Y \subset U$). Since U is clopen, it follows that $X = \overline{Y} \subset U$ as well, so that $U = X$ and thus $V = \varnothing$. $\square$

Example 3.4.14. Let X be the closure of the (path) connected space

$$Y := \left\{ \left(x, \sin\left(\frac{1}{x}\right)\right) : x \in (0, 1] \right\}$$

in $\mathbb{R}^2$. Proposition 3.4.13 immediately yields that X is connected. It is easy to see that

$$X = \left\{ \left(x, \sin\left(\frac{1}{x}\right)\right) : x \in (0, 1] \right\} \cup \{(0, y) : y \in [-1, 1]\}.$$

We claim that there is no path $\gamma \colon [0, 1] \to X$ with $\gamma(0) \in \{0\} \times [-1, 1]$ and $\gamma(1) \in Y$. Assume towards a contradiction that there is such a path, and let $\gamma_1, \gamma_2 \colon [0, 1] \to \mathbb{R}$ be its coordinate functions; that is, $\gamma(t) = (\gamma_1(t), \gamma_2(t))$ for $t \in [0, 1]$. Set $F := \gamma^{-1}(\{0\} \times [-1, 1])$. Then F is closed and thus contains $a := \sup F$. Obviously, $a < 1$ must hold. Replacing $[0, 1]$ with $[a, 1]$, if necessary, we may suppose that $\gamma_1(t) > 0$ for all $t \in (0, 1]$.

Let $\epsilon > 0$ be arbitrary. Since γ_1 is continuous, $\gamma_1([0, \epsilon]) \subset \mathbb{R}$ is connected and thus an interval. Clearly, $\gamma_1([0, \epsilon])$ contains zero and is nondegenerate.

Hence, there is $\delta > 0$ with $[0, \delta] \subset \gamma_1([0, \epsilon])$. Since $\{\sin(\frac{1}{x}) : x \in (0, \delta]\} = [-1, 1]$, and since

$$\gamma_2(t) = \sin\left(\frac{1}{\gamma_1(t)}\right) \qquad (t \in (0, 1]),$$

it follows that $\gamma_2([0, \epsilon]) = [-1, 1]$.

Let $r > 0$ be so small that $[-1, 1] \not\subset [\gamma_2(0) - r, \gamma_2(0) + r]$. The continuity of γ_2 at 0 then yields $\epsilon > 0$ with $\gamma_2([0, \epsilon]) \subset [\gamma_2(0) - r, \gamma_2(0) + r]$, which is impossible if $\gamma([0, \epsilon]) = [-1, 1]$.

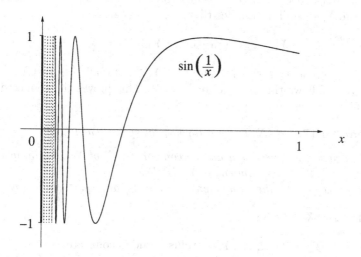

Fig. 3.4: Closure of $\{(x, \sin(\frac{1}{x})) : x \in (0, 1]\}$ in $\mathbb{R}^2$

What if a space is not connected? The few examples of disconnected spaces we have encountered so far—discrete spaces and, for instance, $(1, 2) \cup [7, 8)$—have arisen as disjoint unions of connected "building blocks." We show that this phenomenon occurs generally. First, of course, we have to make precise what we mean by a connected building block.

Definition 3.4.15. *Let $(X, \mathcal{T})$ be a topological space. A* component *of X is a connected subspace of X that is not properly contained in any other connected subspace of X.*

As we show in Theorem 3.4.17 below, *every* topological space can be "broken down" into its components. For its proof, the following proposition is crucial.

Proposition 3.4.16. *Let $(X, \mathcal{T})$ be a topological space, and let $\mathcal{Y}$ be a family of connected subspaces of X such that $Y_1 \cap Y_2 \neq \varnothing$ for any $Y_1, Y_2 \in \mathcal{Y}$. Then $\bigcup\{Y : Y \in \mathcal{Y}\}$ is connected.*

Proof. Set $Y_0 := \bigcup\{Y : Y \in \mathcal{Y}\}$. To see that Y_0 is connected, let $U, V \in \mathcal{T}$ be such that

$$(Y_0 \cap U) \cap (Y_0 \cap V) = \varnothing \quad \text{and} \quad (Y_0 \cap U) \cup (Y_0 \cap V) = Y_0.$$

For each $Y \in \mathcal{Y}$, we thus have

$$(Y \cap U) \cap (Y \cap V) = \varnothing \quad \text{and} \quad (Y \cap U) \cup (Y \cap V) = Y,$$

and since all subspaces $Y \in \mathcal{Y}$ are connected, it follows that each $Y \in \mathcal{Y}$ is either contained in U or in V. Assume that there are $Y_1, Y_2 \in \mathcal{Y}$ such that $Y_1 \subset U$ and $Y_2 \subset V$. This implies that

$$Y_1 \cap Y_2 \subset (Y_0 \cap U) \cap (Y_0 \cap V) = \varnothing,$$

which is impossible. Consequently, we have $Y \subset U$ for all $Y \in \mathcal{Y}$ or $Y \subset V$ for all $Y \in \mathcal{Y}$. It follows that $Y_0 \subset U$ or $Y_0 \subset V$. This proves the connectedness of Y_0. □

Theorem 3.4.17. *Let $(X, \mathcal{T})$ be a topological space. Then:*

(i)   *For each $x \in X$, there is a unique component Y_x of X containing x;*
(ii) *For each $x \in X$, the component Y_x is closed;*
(iii) *For any $x, y \in X$, the components Y_x and Y_y are equal or disjoint.*

Proof. For $x \in X$, let

$$\mathcal{Y}_x := \{Y \subset X : Y \text{ contains } x \text{ and is connected}\}.$$

Note that $\mathcal{Y}_x \neq \varnothing$ because $\{x\} \in \mathcal{Y}_x$. Since $x \in \bigcap\{Y : Y \in \mathcal{Y}_x\}$, Proposition 3.4.16 yields that $Y_x := \bigcup\{Y : Y \in \mathcal{Y}_x\}$ is connected. By its definition, Y_x contains x and is not properly contained in any other connected subspace of X. This proves the existence part of (i) (the uniqueness will follow from (iii)).

For (ii), note that $\overline{Y_x}$ is again connected by Proposition 3.4.13, so that $\overline{Y_x} = Y_x$.

For (iii), let $x, y \in X$ be such that $Y_x \cap Y_y \neq \varnothing$. Then Proposition 3.4.16 yields the connectedness of $Y_x \cup Y_y$, so that $Y_x \cup Y_y = Y_x$ and thus $Y_y \subset Y_x$. Interchanging the rôles of x and y, we obtain $Y_x = Y_y$. □

Loosely speaking, every topological space is the disjoint union of its components.

Examples 3.4.18. (a) If $(X, \mathcal{T})$ is connected, then X is the only component of X.

(b) The components of, say $[-1, 0) \cup [2, 3] \cup (\pi, 7)$, are $[-1, 0)$, $[2, 3]$, and $(\pi, 7)$.

(c) In a discrete space, the components are just the singleton subsets.

(d) We claim that the components of $\mathbb{Q}$, when equipped with the topology inherited from $\mathbb{R}$, are just the singleton subsets (even though $\mathbb{Q}$ is not discrete). Any connected subspace of $\mathbb{Q}$ is also a connected subspace of $\mathbb{R}$ and therefore an interval by Example 3.4.10. Since the only intervals contained in $\mathbb{Q}$ are the degenerate ones (i.e., those consisting of a single point only), the claim follows.

(e) The *Cantor set* C is constructed as follows. Let $C_1 := [0, 1]$, let

$$C_2 := \left[0, \frac{1}{3}\right] \cup \left[\frac{2}{3}, 1\right],$$

let

$$C_3 := \left[0, \frac{1}{9}\right] \cup \left[\frac{2}{9}, \frac{1}{3}\right] \cup \left[\frac{2}{3}, \frac{7}{9}\right] \cup \left[\frac{8}{9}, 1\right],$$

and continue the construction of $C_1 \supset C_2 \supset C_3 \supset \cdots$ as follows. Obtain C_{n+1} from C_n by removing the "middle third" of each of the intervals making up C_n. Define $C := \bigcap_{n=1}^{\infty} C_n$. Clearly, C is a closed subset of $[0, 1]$ and thus compact. It is easy to see that C is infinite: by construction, it contains $\frac{1}{3}, \frac{1}{9}, \frac{1}{27}, \ldots$. (In fact, C is even uncountable; see Exercise 7 below.) For $n \in \mathbb{N}$, let χ_n denote the indicator function of C_n; that is,

$$\chi_n(t) := \begin{cases} 1, & t \in C_n, \\ 0, & \text{otherwise,} \end{cases}$$

and let

$$\mu(C_n) := \int_0^1 \chi_n(t)\, dt$$

be the *Jordan content* of C_n, so that

$$\mu(C_n) = \frac{2}{3}\mu(C_{n-1}) = \cdots = \left(\frac{2}{3}\right)^{n-1},$$

as can easily be seen by induction. Let $a, b \in \mathbb{R}$ be such that $a \leq b$ and $[a, b] \subset C$. Assume that $a < b$, and choose $n \in \mathbb{N}$ so large that $\left(\frac{2}{3}\right)^{n-1} < b - a$. On the other hand, since $[a, b] \subset C_n$, we have

$$b - a \leq \mu(C_n) \leq \left(\frac{2}{3}\right)^{n-1},$$

which contradicts the choice of n. Hence, $a = b$, so that C does not contain any nondegenerate intervals. Consequently, the components of C are its singleton subsets.

The phenomenon displayed in the last three of the preceding examples (not necessarily discrete spaces whose only components are the singleton subsets) warrants another definition.

Definition 3.4.19. *A topological space* $(X, \mathcal{T})$ *is called* totally disconnected *if, for each* $x \in X$, *the component of* X *containing* x *is* $\{x\}$.

Of course, a topological space is totally disconnected if and only if its only connected subsets are the singletons.

At the end of this section, we turn to local versions of both connectedness and path connectedness.

Definition 3.4.20. *A topological space* $(X, \mathcal{T})$ *is called* locally (path) con-nected *if* $\mathcal{N}_x$ *has a base consisting of (path) connected sets for each* $x \in X$.

Example 3.4.21. Open balls in normed spaces are path connected. Hence, any open subset of a normed space is locally path connected.

It is easy to come up with locally (path) connected spaces that fail to be (path) connected, for instance, $(-1, 0) \cup (0, 1)$, but it is less obvious that the converse implication may fail as well.

Example 3.4.22. Let

$$X = \{(x, 0) : x \in [0, 1]\} \cup \{(0, y) : y \in [0, 1]\} \cup \bigcup_{n=1}^{\infty} \left\{ \left(\frac{1}{n}, y \right) : y \in [0, 1] \right\} \subset \mathbb{R}^2.$$

Intuitively, one can think of X as a comb with infinitely many teeth.

We claim that X is path connected. To see this, let $(x_0, y_0), (x_1, y_1) \in X$, and define $\gamma \colon [0, 1] \to \mathbb{R}^2$ by letting

$$\gamma(t) := \begin{cases} (x_0, (1 - 3t)y_0), & t \in \left[0, \frac{1}{3}\right], \\ (x_0 + (3t - 1)(x_1 - x_0), 0), & t \in \left[\frac{1}{3}, \frac{2}{3}\right], \\ (x_1, (3t - 2)y_1), & t \in \left[\frac{2}{3}, 1\right]. \end{cases}$$

It is easy to see that γ is a path in X that connects (x_0, y_0) with (x_1, y_1). To show that X is *not* locally connected (let alone locally path connected), let $y' \in (0, 1]$, let $r \in (0, y')$, and let C contain $(0, y')$ such that $C \subset X \cap B_r((0, y'))$. Since $r < y'$, it is immediate that $C \cap \{(x, 0) : x \in [0, 1]\} = \varnothing$, so that

$$C \subset \{(0, y) : y \in [0, 1]\} \cup \bigcup_{n=1}^{\infty} \left\{ \left(\frac{1}{n}, y \right) : y \in [0, 1] \right\}.$$

This, in turn, yields that

$$\{0\} \subsetneq I \subset \{0\} \cup \left\{ \frac{1}{n} : n \in \mathbb{N} \right\}, \tag{$**$}$$

where I is the image of C under the projection onto the x-axis. Suppose now that C is connected. Since the projection from $\mathbb{R}^2$ onto the x-axis is continu-ous, $I \subset \mathbb{R}$ is also connected and thus an interval. From $(**)$ we conclude that

$I = \{0\}$. Finally, assume that C is a neighborhood of $(0, y')$, so that there is $\epsilon > 0$ with $X \cap B_\epsilon((0, y')) \subset C$. Since $\left(\frac{1}{n}, y'\right) \in X \cap B_\epsilon((0, y'))$ for $n \in \mathbb{N}$ such that $\frac{1}{n} < \epsilon$, there must be nonzero points in I, which is impossible.

All in all, $X \cap B_r((0, y'))$ cannot contain a connected neighborhood of $(0, y')$, so that X is not locally connected.

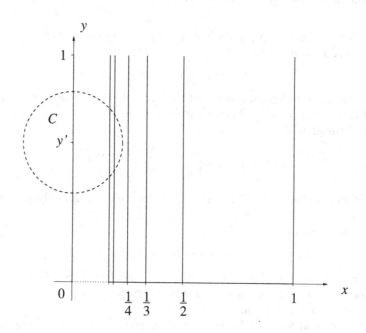

Fig. 3.5: A path connected, but not locally connected space

In a similar vein, one can show that the connected, but not path connected space from Example 3.4.14 is not locally connected (Exercise 12 below).

Here is a reason why local connectedness is of importance.

Proposition 3.4.23. *Let* $(X, \mathcal{T})$ *be a locally connected topological space, and let* U *be an open subspace of* X. *Then the components of* U *are open.*

Proof. Let Y be a component of U, and let $x \in Y$. Since U is open, it is a neighborhood of x in X and thus contains a connected neighborhood, say N, of x. Since $x \in Y \cap N$, Proposition 3.4.16 yields the connectedness of $Y \cup N$. Theorem 3.4.17(i) finally implies that $N \subset Y$, so that Y is a neighborhood of x. Since $x \in Y$ was arbitrary, this proves the openness of Y. $\quad\square$

Corollary 3.4.24. *Let* $(X, \mathcal{T})$ *be a locally connected topological space. Then the components of* X *are open (and thus clopen).*

Local path connectedness combined with connectedness yields path connectedness; in fact, a weaker condition is sufficient.

Proposition 3.4.25. *Let $(X, \mathcal{T})$ be a connected topological space. Then the following are equivalent.*

(i) *X is path connected.*
(ii) *Each point of X has a path connected neighborhood.*

For the proof, we require two constructions on paths that are also handy later on.

- Let γ be any path in a topological space X. Then the *reversed path* γ^{-1}: $[0, 1] \to X$ is defined through

$$\gamma^{-1}(t) := \gamma(1 - t) \qquad (t \in [0, 1]).$$

- Let X be any topological space, and let $\gamma_1, \gamma_2 \colon [0, 1] \to X$ be any two paths such that $\gamma_1(1) = \gamma_2(0)$. Then their *concatenation* $\gamma_1 \odot \gamma_2 \colon [0, 1] \to X$ is defined through

$$(\gamma_1 \odot \gamma_2)(t) := \begin{cases} \gamma_1(2t), & t \in \left[0, \frac{1}{2}\right], \\ \gamma_2(2t - 1), & t \in \left[\frac{1}{2}, 1\right]. \end{cases}$$

Proof (of Proposition 3.4.25). Since $X \in \mathcal{N}_x$ for each $x \in X$, only (ii) $\Longrightarrow$ (i) needs proof.

Fix $x_0 \in X$, and let

$$Y := \{x \in X : x_0 \text{ and } x \text{ can be connected by a path}\}.$$

Clearly, $x_0 \in Y$, so that $Y \neq \varnothing$. It is also easy to see that Y is path connected: for $x_1, x_2 \in Y$, let γ_j be a path connecting x_0 with x_j for $j = 1, 2$. It follows that $\gamma_1^{-1} \odot \gamma_2$ connects x_1 and x_2.

Let $x \in Y$, and let $N \in \mathcal{N}_x$ be path connected. Let $y \in N$. Then x_0 and x can be connected by a path as can x and y. Concatenating the respective paths yields that x_0 and y can be connected by a path. Consequently, $y \in Y$ holds, and since $y \in N$ was arbitrary, this means that $N \subset Y$, so that Y is a neighborhood of x. Since x was arbitrary, it follows that Y is open.

Let $x \in \overline{Y}$, and let $N \in \mathcal{N}_x$ be path connected. Since $x \in \overline{Y}$, there is $y \in N \cap Y$. Then x_0 can be connected with y by a path, and y can be connected with x by a path; concatenation of paths again yields that x_0 and x can be connected by a path. It follows that $x \in Y$, so that $\overline{Y} = Y$.

We have just seen that $Y \neq \varnothing$ is clopen. Since X is connected, this means that $X = Y$, so that X is in fact path connected. $\quad\Box$

Corollary 3.4.26. *Let $(X, \mathcal{T})$ be a connected, locally path connected space. Then X is path connected.*

Examples 3.4.27. (a) Any connected open subset of a normed space is path connected.

(b) The space discussed in Example 3.4.14 is connected, but not path connected and therefore fails to be locally path connected. (In fact, it even fails to be locally connected although this is harder to show; see Exercise 12 below.)

Exercises

1. Let $\mathcal{C}$ be a family of convex subsets of a linear space. Show that $\bigcap\{C : C \in \mathcal{C}\}$ is convex. Is $\bigcup\{C : C \in \mathcal{C}\}$ necessarily convex?

2. Let $((X_i, \mathcal{T}_i))_{i\in\mathbb{I}}$ be a family of topological spaces, and let $(X, \mathcal{T})$ be their topological product. Show that $(X, \mathcal{T})$ is path connected if and only if $(X_i, \mathcal{T}_i)$ is path connected for each $i \in \mathbb{I}$.

3. Show that $\mathbb{R}^n \setminus \{0\}$ is (path) connected if and only if $n \geq 2$, and conclude that $\mathbb{R}^n$ with $n \geq 2$ and $\mathbb{R}$ are not homeomorphic.

4. Let $(X, \mathcal{T}_X)$ and $(Y, \mathcal{T}_Y)$ be connected topological spaces. Show that $X \times Y$ equipped with the product topology is also connected. (*Hint*: For $y \in Y$ fixed, apply Proposition 3.4.16 to the family $\{(X \times \{y\}) \cup (\{x\} \times Y) : x \in X\}$.)

5. Let $((X_i, \mathcal{T}_i))_{i\in\mathbb{I}}$ be a family of topological spaces, and let $(X, \mathcal{T})$ be their topological product. Show that $(X, \mathcal{T})$ is connected if and only if $(X_i, \mathcal{T}_i)$ is connected for each $i \in \mathbb{I}$. For the "if" part, proceed as follows.
   (a) Use Exercise 4 to prove the claim in case $\mathbb{I}$ is finite.
   (b) Fix $(x_i)_{i\in\mathbb{I}} \in X$. For any $\mathbb{J} \subset \mathbb{I}$, let $X_{\mathbb{J}}$ consist of those $(y_i)_{i\in\mathbb{I}} \in X$ such that $y_i = x_i$ for $i \in \mathbb{I} \setminus \mathbb{J}$. Use (a) to prove that $X_{\mathbb{J}}$ is connected whenever $\mathbb{J}$ is finite.
   (c) Use (b) and Proposition 3.4.16 to conclude that $\bigcup\{X_{\mathbb{J}} : \mathbb{J} \subset \mathbb{I} \text{ is finite}\}$ is connected.
   (d) Finally, use (c) and Proposition 3.4.13 to complete the proof that X is connected.

6. Let G be a topological group, that is., a group equipped with a Hausdorff topology such that

$$G \times G \to G, \quad (x, y) \mapsto xy \qquad \text{and} \qquad G \to G, \quad x \mapsto x^{-1}$$

   are continuous, and let G_0 be the component of G containing the identity element. Show that G_0 is a closed normal subgroup of G.

7. Show that $t \in [0, 1]$ lies in the Cantor set C if and only if it has a ternary expansion where all digits that occur are 0 or 2; that is, there is a sequence $(\sigma_n)_{n=1}^\infty$ in $\{0, 2\}$ such that $t = \sum_{n=1}^\infty \frac{\sigma_n}{3^n}$. Conclude that $|C| = \mathfrak{c}$.

8. Let $((X_i, \mathcal{T}_i))_{i\in\mathbb{I}}$ be a family of topological spaces, and let $(X, \mathcal{T})$ be their topological product. Show that $(X, \mathcal{T})$ is totally disconnected if and only if $(X_i, \mathcal{T}_i)$ is totally disconnected for each $i \in \mathbb{I}$.

9. Let $(X, \mathcal{T}_X)$ be a connected space, and let $(Y, \mathcal{T}_Y)$ be a totally disconnected space. Show that every continuous map $f \colon X \to Y$ must be constant.

10. A topological space $(X, \mathcal{T})$ is called *zero-dimensional* if, for any $x, y \in X$ with $x \neq y$, there are $U, V \in \mathcal{T}$ with $x \in U$, $y \in V$, $U \cap V = \varnothing$, and $U \cup V = X$.
   (a) Show that every zero-dimensional space is a totally disconnected Hausdorff space.
   (b) Show that $\mathbb{Q}$ is zero-dimensional (like any totally disconnected subspace of $\mathbb{R}$).
   (c) Let $(X, \mathcal{T})$ be a Hausdorff space such that $\mathcal{T}$ has a base consisting of clopen sets. Show that X is zero-dimensional.
   (d) Let $(K, \mathcal{T})$ be a compact Hausdorff space. Show that K is zero-dimensional if and only if $\mathcal{T}$ has a base consisting of clopen sets. (*Hint*: If K is zero-dimensional, the clopen subsets of K form a base for a Hausdorff topology coarser than $\mathcal{T}$.)

11. Let
$$X := \left\{ \frac{1}{n} : n \in \mathbb{N} \right\} \cup \{0\} \qquad \text{and} \qquad Y := \{0, 1\}$$

have their relative topologies as subspaces of $\mathbb{R}$, and let $X \times Y$ be equipped with the product topology. Define an equivalence relation $\approx$ on $X \times Y$ by letting

$$(x_1, y_1) \approx (x_2, y_2) \quad :\Longleftrightarrow \quad x_1 = x_2 \neq 0.$$

Show that the quotient space $(X \times Y)/\approx$ is totally disconnected, but not Hausdorff (and thus not zero-dimensional).

12. Show that the connected subspace

$$\left\{ \left(x, \sin\left(\frac{1}{x} \right) \right) : x \in (0, 1] \right\} \cup \{ (0, y) : y \in [-1, 1] \}$$

of $\mathbb{R}^2$ is not locally connected.

3.5 Separation Properties

In a metric space, the metric enables us to separate points: any two distinct points have a strictly positive distance. In general topological spaces, separating points from each other is more subtle.

We have already encountered one of the so-called *separation axioms*, namely the Hausdorff separation property (Definition 3.1.3). This section is devoted to the discussion of other separation properties: some stronger, some weaker than the Hausdorff property.

Our first separation axiom is indeed very weak.

Definition 3.5.1. *A topological space* $(X, \mathcal{T})$ *is called a* T_0- *or* Kolmogorov *space if, for any* $x, y \in X$ *with* $x \neq y$, *there is an open set* $U \subset X$ *with* $x \in U$ *and* $y \notin U$ *or* $y \in U$ *and* $x \notin U$.

Examples 3.5.2. (a) Let X be any set with at least two elements equipped with the chaotic topology. Then X is not T_0.

(b) Let R be a commutative ring with identity, and let $\mathfrak{p}, \mathfrak{q} \in \mathrm{Spec}(R)$ be such that $\mathfrak{p} \neq \mathfrak{q}$. Without loss of generality, suppose that $\mathfrak{q} \not\subset \mathfrak{p}$. By the definition of the Zariski topology, the set $U := \mathrm{Spec}(R) \setminus V(\mathfrak{q})$ is open, contains $\mathfrak{p}$, but not $\mathfrak{q}$. Hence, $\mathrm{Spec}(R)$ is T_0.

The next separation axiom is somewhat stronger.

Definition 3.5.3. *A topological space* $(X, \mathcal{T})$ *is called a* T_1-*space if, for any* $x, y \in X$ *with* $x \neq y$, *there are open sets* $U, V \subset X$ *with* $x \in U$, *but* $y \notin U$, *and* $y \in V$, *but* $x \notin V$.

Obviously, every T_1-space is T_0, and every Hausdorff space is T_1. But what about the converse inclusions? The following proposition helps.

Proposition 3.5.4. *Let* $(X, \mathcal{T})$ *be a topological space. Then* X *is a* T_1-*space if and only if* $\{x\}$ *is closed for each* $x \in X$.

Proof. Suppose that X is a T_1-space, and let $x \in X$. For any $y \in X$ with $y \neq x$, there is an open subset U_y of X with $y \in U_y$, but $x \notin U_y$. It follows that

$$X \setminus \{x\} = \bigcup \{U_y : y \in X, \, x \neq y\}$$

is open, so that $\{x\}$ is closed.

Conversely, suppose that all singleton subsets of X are closed, and let $x, y \in X$ be such that $x \neq y$. Then $U := X \setminus \{y\}$ and $V := X \setminus \{x\}$ satisfy the requirements of Definition 3.5.3. $\square$

Examples 3.5.5. (a) Let X be any set, and let $\mathcal{T}$ be the topology consisting of $\varnothing$ and those subsets of X that have a finite complement. Then, trivially, all singleton subsets of X are closed, so that X is a T_1-space by Proposition 3.5.4. However, unless X is finite, the space $(X, \mathcal{T})$ is not Hausdorff (see Example 3.1.4(c)).

(b) Let R be a commutative ring with identity. As we just saw, $\mathrm{Spec}(R)$ is always a T_0-space. By Proposition 3.5.4 and Exercise 3.1.3, $\mathrm{Spec}(R)$ is T_1 if and only if each prime ideal of R is maximal. If R is an integral domain, for instance, but not a field ($R = \mathbb{Z}$, say) then $\mathrm{Spec}(R)$ is a T_0-space that fails to be a T_1-space.

The T in T_0 and T_1 comes from the German word "Trennungsaxiom" (separation axiom). Hausdorff spaces are sometimes called T_2-spaces, and some authors generally label separation properties with a tag of the form T_t, where t is some nonnegative number (and at most five, to my knowledge).

Our next separation axiom has a somewhat different flavor; it is not defined in terms of the topology, but via continuous functions.

Given metric spaces (X, d_X) and (Y, d_Y), we used the symbol $C_b(X, Y)$ to denote the continuous functions in $B(X, Y)$. We continue to use the same symbol if X is merely a topological space.

Definition 3.5.6. *Let $(X, \mathcal{T})$ be a T_1-space. Then X is called* completely regular *if and only if, for each $x \in X$ and each closed set $F \subset X$ with $x \notin F$, there is $f \in C_b(X, \mathbb{R})$ with $f(X) \subset [0,1]$, $f(x) = 1$, and $f|_F = 0$.*

Examples 3.5.7. (a) Let (X, d) be a metric space, let $x_0 \in X$, and let $F \subset X$ be closed such that $x_0 \notin F$. To avoid triviality, suppose that $F \neq \varnothing$. Define

$$g \colon X \to \mathbb{R}, \quad x \mapsto \mathrm{dist}(x, F).$$

Then g is continuous with $g|_F = 0$ and $g(x_0) \neq 0$. Define $f \colon X \to \mathbb{R}$ by letting

$$f(x) := \min \left\{ \frac{g(x)}{g(x_0)}, 1 \right\} \qquad (x \in X).$$

Then $f(X) \subset [0,1]$, $f(x_0) = 1$, and $f|_F = 0$, so that X is completely regular.

(b) The *Sorgenfrey line* is $\mathbb{R}$ equipped with the *Sorgenfrey topology*, that is, the collection of all unions of half-open intervals $[a, b)$ with $a < b$. The Sorgenfrey topology is finer than the canonical topology because

$$(a, b) = \bigcup \{[c, b) : a < c < b\}$$

is Sorgenfrey open for any $a < b$. Let $a \in \mathbb{R}$; then the half line

$$[a, \infty) = \bigcup \{[a, b) : b \in \mathbb{R}, \, b > a\}$$

is Sorgenfrey open, and since it is closed with respect to the canonical topology on $\mathbb{R}$, it is also Sorgenfrey closed. Consequently, whenever $a < b$, the half-open interval

$$[a, b) = [a, \infty) \setminus [b, \infty)$$

is Sorgenfrey clopen. Let $F \subset \mathbb{R}$ be Sorgenfrey closed, and let $x \in \mathbb{R} \setminus F$. By the definition of the Sorgenfrey topology, there are $a < b$ such that $x \in [a, b) \subset \mathbb{R} \setminus F$. Let f be the indicator function of $[a, b)$. Since $[a, b)$ is Sorgenfrey clopen, f is continuous (and clearly satisfies $f(\mathbb{R}) \subset [0, 1]$, $f(x) = 1$, and $f|_F = 0$). Consequently, the Sorgenfrey line is completely regular (even though it is not metrizable; see Example 3.5.14 below).

(c) Every subspace of a completely regular space is again completely regular.

How does complete regularity relate to the separation axioms we have encountered so far? One implication is fairly straightforward.

Proposition 3.5.8. *Let $(X, \mathcal{T})$ be a completely regular space, let $x \in X$, and let $F \subset X$ be closed such that $x \notin F$. Then there are open subsets U and V of X such that $x \in U$, $F \subset V$, and $U \cap V = \varnothing$. In particular, X is Hausdorff.*

Proof. Let $f \colon X \to \mathbb{R}$ be continuous such that $f(x) = 1$ and $f|_F = 0$. Let

$$U := \left\{ y \in X : f(y) > \frac{1}{2} \right\} \qquad \text{and} \qquad V := \left\{ y \in X : f(y) < \frac{1}{2} \right\}.$$

It follows that U and V are open, that $x \in U$ and $F \subset V$, and that $U \cap V = \varnothing$.

Singleton subsets of a T_1-space are closed, thus it is immediate that X is Hausdorff. □

Is, perhaps, complete regularity equivalent to the Hausdorff separation property? The following example gives the (negative) answer.

Example 3.5.9. For each $x \in \mathbb{R}$, we define a system $\mathfrak{N}_x$ of neighborhoods as follows. If $x \neq 0$, let $\mathfrak{N}_x$ be the system of neighborhoods of x in the ordinary topology; if $x = 0$, let $\mathfrak{N}_x$ consist of all sets containing a set of the form

$$(-\epsilon, \epsilon) \setminus \left\{ \frac{1}{n} : n \in \mathbb{N} \right\}$$

with $\epsilon > 0$. By Theorem 3.1.10, there is a unique topology $\mathcal{T}$ on $\mathbb{R}$ such that $\mathfrak{N}_x = \mathcal{N}_x$ for $x \in \mathbb{R}$. Clearly, $\mathcal{T}$ is finer than the usual topology on $\mathbb{R}$ and thus Hausdorff.

Let

$$F = \left\{ \frac{1}{n} : n \in \mathbb{N} \right\}.$$

The way $\mathcal{T}$ is defined, it is obvious that $\mathbb{R} \setminus F$ is a neighborhood of each of its points, so that F is closed. Assume that $(\mathbb{R}, \mathcal{T})$ is completely regular. By Proposition 3.5.8, there are $U, V \in \mathcal{T}$ such that $0 \in U$, $F \subset V$, and $U \cap V = \varnothing$. By the definition of $\mathcal{N}_0$, there is $\epsilon > 0$ such that

$$(-\epsilon, \epsilon) \setminus \left\{ \frac{1}{n} : n \in \mathbb{N} \right\} \subset U,$$

and by the definition of $\mathcal{N}_x$ for $x \neq 0$, there is, for each $n \in \mathbb{N}$, a number $\epsilon_n > 0$ such that

$$\left(\frac{1}{n} - \epsilon_n, \frac{1}{n} + \epsilon_n \right) \subset V.$$

Since $U \cap V = \varnothing$, it follows that

$$\left((-\epsilon, \epsilon) \setminus \left\{ \frac{1}{n} : n \in \mathbb{N} \right\} \right) \cap \left(\frac{1}{n} - \epsilon_n, \frac{1}{n} + \epsilon_n \right) = \varnothing \qquad (n \in \mathbb{N}).$$

This, however, is impossible for $\frac{1}{n} < \epsilon$. (Incidentally, this example is first countable, but not metrizable because otherwise it would be completely regular.)

We conclude our discussion of separation axioms with yet another one.

Definition 3.5.10. *A T_1-space $(X, \mathcal{T})$ is called* normal *if, for any closed sets $F, G \subset X$ with $F \cap G = \varnothing$, there are open sets $U, V \subset X$ with $F \subset U$, $G \subset V$, and $U \cap V = \varnothing$.*

Examples 3.5.11. (a) Let (X, d) be a metric space, and let $F, G \subset X$ be closed (and nonempty, to avoid triviality). Then

$$f : X \to \mathbb{R}, \quad x \mapsto \mathrm{dist}(x, F) - \mathrm{dist}(x, G)$$

is continuous. Let

$$U := \{ x \in X : f(x) < 0 \} \quad \text{and} \quad V := \{ x \in X : f(x) > 0 \}.$$

Then U and V are open such that $F \subset U$, $G \subset V$, and $U \cap V = \varnothing$. Hence, X is normal.

(b) Let $(K, \mathcal{T})$ be a compact Hausdorff space, and let $F, G \subset K$ be closed and disjoint (and again nonempty, to avoid triviality). Fix $x \in F$. For $y \in G$, there are $U_y, V_y \in \mathcal{T}$ such that $x \in U_y$, $y \in V_y$, $U_y \cap V_y = \varnothing$. Obviously,

$\{V_y : y \in G\}$ is an open cover for G, and since G (being a closed subspace of a compact space) is compact, there are $y_1, \ldots, y_n \in Y$ such that

$$G \subset V_{y_1} \cup \cdots \cup V_{y_n}.$$

Let

$$U_x := U_{y_1} \cap \cdots \cap U_{y_n} \qquad \text{and} \qquad V_x := V_{y_1} \cup \cdots \cup V_{y_n}.$$

Then U_x and V_x are open such that $x \in U_x$, $G \subset V_x$, and $U_x \cap V_x = \varnothing$. Clearly $\{U_x : x \in F\}$ is an open cover for F, and therefore, there are $x_1, \ldots, x_m \in F$ such that

$$F \subset U_{x_1} \cup \cdots \cup U_{x_m}.$$

Letting

$$U := U_{x_1} \cup \cdots \cup U_{x_m} \qquad \text{and} \qquad V := V_{x_1} \cap \cdots \cap V_{x_m},$$

we obtain open subsets of X with $F \subset U$, $G \subset V$, and $U \cap V = \varnothing$. All in all, K is normal.

All normal spaces are trivially Hausdorff, and Example 3.5.9 shows that the converse is false. But what is the relation between normality and complete regularity? We give a complete answer in the next chapter. For now, we content ourselves with giving an example of a completely regular space that fails to be normal.

We first prove an elementary hereditary property for complete regularity.

Proposition 3.5.12. *Let $((X_i, \mathcal{T}_i))_{i \in \mathbb{I}}$ be a family of completely regular spaces, and let $(X, \mathcal{T})$ denote their topological product. Then X is completely regular.*

Proof. For $i \in \mathbb{I}$, we use $\pi_i : X \to X_i$ as usual to denote the ith coordinate projection.

Let $x = (x_i)_{i \in \mathbb{I}}$ and $y = (y_i)_{i \in \mathbb{I}}$ be distinct points of X. Hence, there is $i_0 \in \mathbb{I}$ such that $x_{i_0} \neq y_{i_0}$. Since X_{i_0} is a T_1-space, there are $U_{i_0}, V_{i_0} \in \mathcal{T}_{i_0}$ with $x_{i_0} \in U_{i_0}$, $y_{i_0} \notin U_{i_0}$, $y_{i_0} \in V_{i_0}$, and $x_{i_0} \notin V_{i_0}$. Let $U := \pi_{i_0}^{-1}(U_{i_0})$ and $V := \pi_{i_0}^{-1}(V_{i_0})$. Then $U, V \subset X$ are open such that $x \in U$, $y \notin U$, $y \in V$, and $x \notin V$. Hence, X is a T_1-space.

Let $\varnothing \neq F \subset X$ be closed, and let $x = (x_i)_{i \in \mathbb{I}} \in X \setminus F$. By the definition of the product topology, there are $i_1, \ldots, i_n \in \mathbb{I}$ and $U_j \in \mathcal{T}_{i_j}$ for $j = 1, \ldots, n$ with

$$x \in \pi_{i_1}^{-1}(U_{i_1}) \cap \cdots \cap \pi_{i_n}^{-1}(U_{i_n}) \subset X \setminus F.$$

The spaces $X_{i_1}, \ldots, X_{i_n}$ are all completely regular. Hence, for $j = 1, \ldots, n$, there is $f_j \in C_b(X_{i_j}, \mathbb{R})$ with $f_j(X_{i_j}) \subset [0, 1]$ such that

$$f_j(x_{i_j}) = 1 \qquad \text{and} \qquad f|_{X_{i_j} \setminus U_j} = 0.$$

Define

$$f: X \to \mathbb{R}, \quad y \mapsto \prod_{j=1}^{n} f_j(\pi_{i_j}(y)).$$

Then f is continuous with $f(X) \subset [0,1]$, $f(x) = 1$, and $f|_F = 0$. All in all, X is completely regular. $\square$

Next, we prove a somewhat surprising statement about discrete subspaces of normal spaces.

Theorem 3.5.13. *Let $(X, \mathcal{T})$ be a separable normal space, and let D be a closed discrete subspace of X. Then D cannot have cardinality $\mathfrak{c}$ or larger.*

Proof. Let $S \subset D$. Then S and $D \setminus S$ are closed in D (D is discrete!), and since D is closed in X, both S and $D \setminus S$ are also closed in X. Hence, by the normality of X, there are $U_S, V_S \in \mathcal{T}$ with $S \subset U_S$, $D \setminus S \subset V_S$, and $U_S \cap V_S = \varnothing$.

Let C be a dense countable subset of X, and define

$$f: \mathfrak{P}(D) \to \mathfrak{P}(C), \quad S \mapsto C \cap U_S.$$

We claim that f is injective. Let $S, T \subset D$ be such that $S \neq T$. We can suppose that $S \setminus T \neq \varnothing$. It follows that $U_S \cap V_T$ is nonempty (and open). Since C is dense in X, we conclude that $C \cap U_S \cap V_T \neq \varnothing$. On the other hand, $C \cap U_T \cap V_T = \varnothing$ must hold due to the choice of U_T and V_T. Consequently, we have $C \cap U_S \neq C \cap U_T$.

The injectivity of f yields

$$|\mathfrak{P}(D)| \leq |\mathfrak{P}(C)| \leq 2^{\aleph_0} = \mathfrak{c},$$

and since $|D| < |\mathfrak{P}(D)|$, the claim follows. $\square$

Example 3.5.14. Let $(X, \mathcal{T})$ be the *Sorgenfrey plane*, that is, the topological product of the Sorgenfrey line with itself. By Example 3.5.7(b) and Proposition 3.5.12, X is completely regular.

Since the sets of the form

$$[a, b) \times [c, d) \quad (a < b, \, c < d)$$

form a base for $\mathcal{T}$, and since every such set has a nonempty intersection with $\mathbb{Q}^2$, it is clear that $(X, \mathcal{T})$ is separable, and equally clearly, $\mathcal{T}$ is finer than the usual topology on $\mathbb{R}^2$. Since

$$D := \{(x, -x) : x \in \mathbb{R}\}$$

is closed in $\mathbb{R}^2$ with respect to the ordinary topology, it is therefore also closed with respect to $\mathcal{T}$. For $x \in \mathbb{R}$ and $a, b \in \mathbb{R}$ with $x < a$ and $-x < b$, note that

$$D \cap ([x, a) \times [-x, b)) = \{(x, -x)\},$$

so that $(D, \mathcal{T}|_D)$ is discrete (compare Example 3.1.26). Obviously, D has cardinality $\mathfrak{c}$, which, by Theorem 3.5.13, would be impossible if X were normal. (This example also shows that the Sorgenfrey line cannot be metrizable; otherwise the Sorgenfrey plane would be metrizable, too, and thus normal.)

Exercises

1. Show that a topological space $(X, \mathcal{T})$ is T_1 if and only if each constant net in X has a unique limit.
2. Let $(X, \mathcal{T})$ be a T_1-space such that $\mathcal{T}$ has a base consisting of clopen sets. Show that X is completely regular.
3. Let $((X_i, \mathcal{T}_i))_{i \in \mathbb{I}}$ be a family of topological spaces such that their topological product is completely regular. Show that X_i is completely regular for each $i \in \mathbb{I}$.
4. Let $(X, \mathcal{T})$ be a completely regular space with infinitely many points. Show that there is a sequence $(U_n)_{n=1}^\infty$ of nonempty open subsets of X such that $U_n \cap U_m = \varnothing$ for $n \neq m$. Conclude that X contains a countably infinite, discrete subspace. (*Hint*: Show first that there is a non-empty open subset U of X such that $X \setminus \overline{U}$ is infinite.)
5. A topological space $(X, \mathcal{T})$ is called *Lindelöf* if, for each open cover $\mathcal{U}$ of X, there are $U_1, U_2, \ldots \in \mathcal{U}$ with $X = \bigcup_{n=1}^\infty U_n$; that is, every open cover has a countable subcover.
   (a) Suppose that X is σ-*compact*; that is, there is a sequence $(K_n)_{n=1}^\infty$ of compact subsets of X with $X = \bigcup_{n=1}^\infty K_n$. Show that X is Lindelöf.
   (b) Let $\mathcal{B}$ be a base for $\mathcal{T}$. Show that X is Lindelöf if and only if every open cover $\mathcal{U} \subset \mathcal{B}$ of X has a countable subcover (so that, in particular, every second countable space is Lindelöf).
   (c) Show that a closed subspace of a Lindelöf space is again Lindelöf.
6. Let $(X, \mathcal{T})$ be a completely regular Lindelöf space, and let $F, G \subset X$ be closed and disjoint.
   (a) Show that, for each $x \in F$ and $y \in G$, there are open subsets U_x and V_y of X with $x \in U_x$, $y \in V_y$, $G \cap \overline{U}_x = \varnothing$, and $F \cap \overline{V}_y = \varnothing$.
   (b) Argue that there are sequences $(x_n)_{n=1}^\infty$ in F and $(y_n)_{n=1}^\infty$ in G with $F \subset \bigcup_{n=1}^\infty U_{x_n}$ and $G \subset \bigcup_{n=1}^\infty V_{y_n}$.
   (c) Let
   $$U := \bigcup_{n=1}^\infty U_{x_n} \setminus \left(\overline{V}_{y_1} \cup \cdots \cup \overline{V}_{y_n} \right)$$
   and
   $$V := \bigcup_{n=1}^\infty V_{y_n} \setminus \left(\overline{U}_{x_1} \cup \cdots \cup \overline{U}_{x_n} \right).$$
   Show that U and V are open and disjoint with $F \subset U$ and $G \subset V$, and conclude that X is normal.
7. Let $(X, \mathcal{T})$ be the Sorgenfrey line. Show that X is Lindelöf (and thus normal by the previous problem). Proceed as follows.
   (a) Let $\mathcal{U}$ be an open cover for X. Argue that one can suppose without loss of generality that $\mathcal{U}$ only consists of sets of the form $[a, b)$ with $a < b$.
   (b) Let $\mathcal{V} := \{(a, b) : [a, b) \in \mathcal{U}\}$, and let $C := X \setminus \bigcup \{V : V \in \mathcal{V}\}$. Prove that C is countable.
   (c) Argue that $\mathbb{R} \setminus C$ is Lindelöf with respect to the canonical topology on $\mathbb{R}$, and use this and (b) to show that $\mathcal{U}$ has a countable subcover.
   What can you conclude about the normality of the topological product of a family of normal spaces?
8. Show that a closed subspace of a normal space is normal again.

Remarks

In 1895, the French mathematician Henri Poincaré published a book named *Analysis situs* (Latin for *analysis of places*), which is considered the first attempt to systematically study the phenomena that would later be called *topology* (which is derived from the Greek words *topos*, meaning *place*, and *logos*, meaning *study*, and therefore means *study of places*). Further attempts were soon made by Fréchet [FRÉCHET 06] and others.

It was Felix Hausdorff, the one after whom the Hausdorff spaces are named, in [HAUSDORFF 14] who came up with the modern definition along with the modern name (in German, of course). He used the approach presented in Theorem 3.1.10, that is, through an axiomatization of the notion of neighborhood. Kuratowski closure operations, which provide an alternative, but equivalent approach to topology, were introduced by and named after the Polish mathematician Kazimierz Kuratowski in the early 1920s. The approach we give, through an axiomatization of openness, is the most widespread one these days.

Modern introductions to set theoretic topology are, among many others, the books by John L. Kelley [KELLEY 55], George F. Simmons [SIMMONS 63], Stephen Willard [WILLARD 70], Graham J. O. Jameson [JAMESON 74], and James R. Munkres [MUNKRES 00].

The product topology was discovered by the young Andrey N. Tychonoff (Andrei N. Tikhonov), then at most in his early twenties, about 1926. Interestingly, his teacher Pavel S. Alexandroff (Alexandrov) was doubtful if it was a good concept at all. It was, and Tychonoff used it to prove the famous theorem that is now named after him. Nowadays, various proofs of Tychonoff's theorem are available; the one we present is due to Paul R. Chernoff [CHERNOFF 92]. Since Tychonoff's theorem is about Cartesian products, objects whose very existence cannot be guaranteed other than by Zorn's lemma (or one of its equivalent formulations), it isn't much of a surprise that each of its proofs relies on Zorn's lemma in some form. Interestingly, Tychonoff's theorem is not only implied by Zorn's lemma, but equivalent to it [KELLEY 50].

Our definition of total disconnectedness is considered to be the "standard" one, in the sense that most textbooks nowadays use it. But there are exceptions: for example, [SIMMONS 63] calls a space totally disconnected when we call it zero-dimensional as defined in Exercise 3.4.10.

The names for the separation axioms are also not entirely standardized. For instance, in [KELLEY 55], completely regular and normal spaces need not be T_1, and what we call a completely regular space is called a Tychonoff space in [KELLEY 55]. The term T_1 originates in Alexandroff's fundamental treatise [ALEXANDROFF & HOPF 35] with Heinz Hopf: there, the authors consider five separation axioms T_j for $j = 1, \ldots, 5$. The T_2-axiom is nowadays called the Hausdorff separation property, and besides the T_1-axiom, their nomenclature has not survived. Other authors have considered separation axioms labeled T_t with values for t other than $1, \ldots, 5$: this is how the T_0-axiom came into existence (completely regular spaces, for instance, are sometimes—half mockingly—referred to as $T_{3\frac{1}{2}}$-spaces). The separation property considered in Proposition 3.5.8, which is implied by complete regularity and implies the Hausdorff property, is called *regularity*. The subsequent example is of a space which is Hausdorff, but not regular. It is not obvious that regularity is indeed weaker than complete regularity, but it is true; an example, not for the faint of heart, is given as an exercise in [WILLARD 70].

Kuratowski, Tychonoff, and Alexandroff all had long and successful professional careers despite trying political circumstances in their countries, and all died in their eighties. Tychonoff, born in 1906, before the Soviet Union existed, passed away in 1993, when it no longer existed.

Twentieth-century politics wouldn't allow Felix Hausdorff to die in peace. Being Jewish, he was notified in 1942 of his impending deportation to the concentration camp Theresienstadt (Terezín) in occupied Czechoslovakia. Not willing to give up their dignity as human beings, whatever was left of it in 1942, he, his wife, and his sister-in-law took their own lives.

4

Systems of Continuous Functions

Topological spaces were introduced in the first place because they are the natural habitat for continuous functions.

Given two topological spaces X and Y, the number of continuous functions from X to Y can vary greatly, depending on the topologies involved: everything is possible from "all functions are continuous" to "only the constants are continuous." In this chapter, we are interested in continuous functions from topological spaces into $\mathbb{R}$ or $\mathbb{C}$. On a metric space, the metric itself easily provides a plentiful supply of such functions. But can anything meaningful be said in the absence of a metric?

4.1 Urysohn's Lemma and Applications

Let $(X, \mathcal{T})$ be a topological space. Is there any continuous function from X to $\mathbb{R}$ that is not constant?

If X is normal, the surprising (and surprisingly easy to prove) answer is given by Urysohn's lemma, for whose proof we require the following.

Lemma 4.1.1. *Let $(X, \mathcal{T})$ be normal, let $F \subset X$ be closed, and let $U \subset X$ be open such that $F \subset U$. Then there is an open subset V of X such that*

$$F \subset V \subset \overline{V} \subset U.$$

Proof. Since F and $X \setminus U$ are both closed and disjoint, the definition of normality yields disjoint open sets $V, W \subset X$ such that $F \subset V$ and $X \setminus U \subset W$. Since $V \cap W = \varnothing$ (i.e., $V \subset X \setminus W$), and since $X \setminus W$ is closed, it follows that

$$\overline{V} \subset X \setminus W \subset (X \setminus (X \setminus U)) = U,$$

as claimed. $\square$

Theorem 4.1.2 (Urysohn's lemma). *Let $(X, \mathcal{T})$ be a normal topological space, and let F and G be disjoint closed subsets of X. Then there is a continuous function $f \colon X \to \mathbb{R}$ such that $f(X) \subset [0, 1]$, $f|_F = 0$, and $f|_G = 1$.*

Proof. Since $X \setminus G$ is open and contains F, Lemma 4.1.1 yields an open set $U_{\frac{1}{2}} \subset X$ with

$$F \subset U_{\frac{1}{2}} \subset \overline{U}_{\frac{1}{2}} \subset X \setminus G.$$

Via the same argument, we obtain open subsets $U_{\frac{1}{4}}$ and $U_{\frac{3}{4}}$ of X such that

$$F \subset U_{\frac{1}{4}} \subset \overline{U}_{\frac{1}{4}} \subset U_{\frac{1}{2}} \subset \overline{U}_{\frac{1}{2}} \subset U_{\frac{3}{4}} \subset \overline{U}_{\frac{3}{4}} \subset X \setminus G.$$

In the next step, we obtain open sets $U_{\frac{1}{8}}$, $U_{\frac{3}{8}}$, $U_{\frac{5}{8}}$, and $U_{\frac{7}{8}}$ in X, such that

$$F \subset U_{\frac{1}{8}} \subset \overline{U}_{\frac{1}{8}} \subset U_{\frac{1}{4}} \subset \overline{U}_{\frac{1}{4}}$$
$$\subset U_{\frac{3}{8}} \subset \overline{U}_{\frac{3}{8}} \subset U_{\frac{1}{2}} \subset \overline{U}_{\frac{1}{2}} \subset U_{\frac{5}{8}} \subset \overline{U}_{\frac{5}{8}} \subset U_{\frac{3}{4}} \subset \overline{U}_{\frac{3}{4}} \subset U_{\frac{7}{8}} \subset \overline{U}_{\frac{7}{8}} \subset X \setminus G.$$

Let D denote the set of dyadic rationals in $(0,1)$, that is, all numbers of the form $\frac{m}{2^n}$, where $n \in \mathbb{N}$, and $m \in \{1, 2, \ldots, 2^n - 1\}$. Continuing the process outlined before, we obtain, for each $t \in D$, an open subset U_t of X such that, for any $s, t \in D$, with $s < t$, we have

$$F \subset U_s \subset \overline{U}_s \subset U_t \subset \overline{U}_t \subset X \setminus G.$$

Define $f : X \to \mathbb{R}$ as follows,

$$f(x) := \begin{cases} \sup\{t \in D : x \notin U_t\}, & x \notin \bigcap_{t \in D} U_t, \\ 0, & \text{otherwise.} \end{cases}$$

It is clear that $f(X) \subset [0,1]$, that $f|_F = 0$, and that $f|_G = 1$. All that remains to be shown is the continuity of f.

Since $f(X) \subset [0,1]$, it is sufficient to show that $f^{-1}(U)$ is open for each subset U of $[0,1]$ that is open in $[0,1]$, that is, open with respect to the relative topology of $[0,1]$. Since $\{[0,a), (b,1] : a, b \in [0,1]\}$ is a subbase for this topology, it is sufficient to show that $f^{-1}([0,a))$ and $f^{-1}((a,1])$ are open for each $a \in [0,1]$.

Let $a \in [0,1]$. It follows from the definition of f that $f(x) < a$ if and only if there is $t \in D$, $t < a$, with $x \in U_t$. Consequently,

$$f^{-1}([0,a)) = \bigcup_{t < a} U_t$$

is open. Also, $f(x) > a$ holds if and only if there is $t \in D$, $t > a$, with $x \notin \overline{U}_t$. Therefore,

$$f^{-1}((a,1]) = \bigcup_{t > a} X \setminus \overline{U}_t$$

is also open.

It follows that f is indeed continuous. $\square$

It is easy to see that the interval $[0,1]$ in Urysohn's lemma can be replaced by any interval $[a, b]$ with $a < b$.

Corollary 4.1.3. *Let $(X, \mathcal{T})$ be a normal topological space, let F and G be disjoint closed subsets of X, and let $a < b$. Then there is a continuous function $f\colon X \to \mathbb{R}$ such that $f(X) \subset [a, b]$, $f|_F = a$, and $f|_G = b$.*

Proof. By Urysohn's lemma, there is a continuous function $g\colon X \to [0, 1]$ with $g|_F = 0$ and $g|_G = 1$. Define $f\colon X \to [a, b]$ via

$$f(x) := (b - a)g(x) + a \qquad (x \in X),$$

which proves the corollary. $\square$

As another straightforward consequence of Urysohn's lemma, we can now clarify the relation between normality and complete regularity:

Corollary 4.1.4. *Let $(X, \mathcal{T})$ be a normal. Then X is completely regular.*

Further consequences are as follows.

Corollary 4.1.5. *Let $(X, \mathcal{T})$ be a locally compact Hausdorff space. Then X is completely regular.*

Proof. The one-point compactification of X is a compact Hausdorff space, therefore normal, and thus completely regular. As a subspace of a completely regular space, X itself is completely regular. $\square$

Corollary 4.1.6. *The following are equivalent for a topological space $(X, \mathcal{T})$.*

(i) *X is a compact Hausdorff space.*
(ii) *There is an index set $\mathbb{I}$, such that X is homeomorphic to a closed subspace of $[0, 1]^{\mathbb{I}}$.*

Proof. (i) $\Longrightarrow$ (ii): Let

$$\mathbb{I} := \{f \in C(X, \mathbb{R}) : f(X) \subset [0, 1]\},$$

and define

$$\iota\colon X \to [0, 1]^{\mathbb{I}}, \quad x \mapsto (f(x))_{f \in \mathbb{I}}.$$

Clearly, ι is continuous, and by Urysohn's lemma, it is also injective. Hence, $\iota\colon X \to \iota(X)$ is a continuous bijection from a compact space into a Hausdorff space. By Theorem 3.3.11, this means that ι is actually a homeomorphism between X and $\iota(X)$. Finally, since X is compact, so is $\iota(X)$, which is therefore closed in the Hausdorff space $[0, 1]^{\mathbb{I}}$.

(ii) $\Longrightarrow$ (i): Since $[0, 1]^{\mathbb{I}}$ is a compact Hausdorff space, so are each of its closed subspaces. $\square$

Next, we put Urysohn's lemma to work on the question of metrizability of topological spaces. We have already encountered a few necessary properties for a topological space to be metrizable: the Hausdorff separation property, first countability, and normality. None of these properties, however, is sufficient.

We first prove a technical lemma.

Lemma 4.1.7. *Let $(X, \mathcal{T})$ be a topological space, let $\mathcal{B}$ a base for $\mathcal{T}$, and suppose that, for each $U \in \mathcal{B}$ and each $x \in U$, there is a continuous function $f_{U,x} : X \to \mathbb{R}$ such that $f_{U,x}(x) = 1$ and $f_{U,x}|_{X \setminus U} = 0$. Then $\mathcal{T}$ is the coarsest topology on X such that all functions in $\{f_{U,x} : U \in \mathcal{B}, x \in U\}$ are continuous.*

Proof. Let $\mathcal{T}'$ denote the coarsest topology on X such that all functions in $\{f_{U,x} : U \in \mathcal{B}, x \in U\}$ are continuous. Clearly, $\mathcal{T}$ is finer than $\mathcal{T}'$. Assume towards a contradiction that there is a set $U \in \mathcal{T}$ that is not in $\mathcal{T}'$. Without loss of generality, we may suppose that $U \in \mathcal{B}$. If U is not in $\mathcal{T}'$, then $X \setminus U$ is not closed with respect to $\mathcal{T}'$. Consequently, there is $x \notin X \setminus U$ (i.e., $x \in U$) that lies in the closure of $X \setminus U$ with respect to $\mathcal{T}'$. Let $(x_\alpha)_\alpha$ be a net in $X \setminus U$ such that $x_\alpha \xrightarrow{\mathcal{T}'} x$. It follows that

$$1 = f_{U,x}(x) = \lim_\alpha f_{U,x}(x_\alpha) = 0,$$

which is absurd. Hence, $\mathcal{T}$ and $\mathcal{T}'$ must coincide. $\square$

Corollary 4.1.8. *Let $(X, \mathcal{T})$ be a completely regular space. Then $\mathcal{T}$ is the coarsest topology on X such that all functions in $C_b(X, \mathbb{R})$ are continuous.*

The following definition was already given in Exercise 3.1.4, but repeating it won't hurt.

Definition 4.1.9. *A topological space $(X, \mathcal{T})$ is called second countable if $\mathcal{T}$ has a countable base.*

Theorem 4.1.10 (Urysohn's metrization theorem). *Let $(X, \mathcal{T})$ be a normal, second countable space. Then X is metrizable.*

Proof. Let $\{U_1, U_2, U_3, \ldots\} \subset \mathcal{T}$ be a countable base for $\mathcal{T}$, and let

$$\mathbb{A} := \left\{(n, m) \in \mathbb{N}^2 : \overline{U}_n \subset U_m\right\}.$$

We claim that, for each $m \in \mathbb{N}$, and for each $x \in U_m$, there is $n \in \mathbb{N}$ with $x \in U_n \subset \overline{U}_n \subset U_m$ (so that, in particular, (n, m) is in $\mathbb{A}$). Indeed, Lemma 4.1.1 yields $U \in \mathcal{T}$ with $x \in U \subset \overline{U} \subset U_m$. Since $\{U_1, U_2, \ldots\}$ is a base for $\mathcal{T}$, the existence of n as required follows.

By Urysohn's lemma, there is, for each $(n, m) \in \mathbb{A}$, a continuous function $f_{n,m} : X \to \mathbb{R}$ with $f_{n,m}(X) \subset [0, 1]$, $f_{n,m}|_{\overline{U}_n} = 1$, and $f_{n,m}|_{X \setminus U_m} = 0$. Define $d : X \times X \to \mathbb{R}$ by letting

$$d(x, y) := \sum_{(n,m) \in \mathbb{A}} \frac{1}{2^{n+m}} |f_{n,m}(x) - f_{n,m}(y)| \qquad (x, y \in X).$$

It is straightforward that d is a semimetric on X. To see that d is, in fact, a metric, let $x, y \in X$ be such that $x \neq y$. Since X is Hausdorff, there is $m \in \mathbb{N}$ with $x \in U_m$ and $y \notin U_m$. By the foregoing, there is $n \in \mathbb{N}$ such that

$x \in \overline{U}_n \subset U_m$, so that $(n, m) \in \mathbb{A}$. Since $f_{n,m}|_{\overline{U}_n} = 1$, and $f_{n,m}|_{X \setminus U_m} = 0$, we have $|f_{n,m}(x) - f_{n,m}(y)| = |f_{n,m}(x)| = 1$ and thus $d(x, y) \geq \frac{1}{2^{n+m}} > 0$.

It is routine to verify that the identity from $(X, \mathcal{T})$ to (X, d) is continuous—because each function $f_{n,m}$ is continuous—so that the topology induced by d is coarser than $\mathcal{T}$. On the other hand, each function $f_{n,m}$ with $(n, m) \in \mathbb{A}$ is continuous with respect to d, and from Lemma 4.1.7, we conclude that $\mathcal{T}$ is also coarser than the topology of (X, d). $\square$

Corollary 4.1.11. *The following are equivalent for a compact Hausdorff space $(K, \mathcal{T})$.*

(i) *K is second countable.*
(ii) *K is metrizable.*

Proof. (i) $\implies$ (ii) follows immediately from the metrization theorem.

For the converse implication, note that a compact metrizable space is always separable and thus second countable by Exercise 3.1.4. $\square$

The following lemma has already been proven for metric spaces as (a special case of) Example 2.4.6; the proof, however, given there works for general topological spaces as well.

Lemma 4.1.12. *Let $(X, \mathcal{T})$ be a topological space. Then $C_b(X, \mathbb{F})$, equipped with the norm $\| \cdot \|_\infty$ given by*

$$\|f\|_\infty := \sup\{|f(x)| : x \in X\} \qquad (f \in C_b(X, \mathbb{F})),$$

is a Banach space.

We require Lemma 4.1.12 in the proof of yet another application of Urysohn's lemma.

Theorem 4.1.13 (Tietze's extension theorem). *Let $(X, \mathcal{T})$ be a normal space, let Y be a closed subspace, and let $f : Y \to \mathbb{R}$ be continuous such that $f(Y) \subset [a, b]$. Then there is a continuous function $\tilde{f} : X \to \mathbb{R}$ with $\tilde{f}(X) \subset [a, b]$ that extends f.*

Proof. The claim is trivial if $a = b$, so suppose that $a < b$. Without loss of generality, suppose further that $a = -1$ and $b = 1$.

Set $f_0 := f$, and let

$$F_0 := \left\{ x \in Y : f_0(x) \leq -\frac{1}{3} \right\} \qquad \text{and} \qquad G_0 := \left\{ x \in Y : f_0(x) \geq \frac{1}{3} \right\}.$$

Then F_0 and G_0 are closed and disjoint. By Corollary 4.1.3, there is a continuous function $g_0 : X \to [-\frac{1}{3}, \frac{1}{3}]$ such that $g_0|_{F_0} = -\frac{1}{3}$ and $g_0|_{G_0} = \frac{1}{3}$. Let $f_1 := f_0 - g_0|_Y$. It follows that $f_1(Y) \subset [-\frac{2}{3}, \frac{2}{3}]$. Let

$$F_1 := \left\{ x \in Y : f_1(x) \leq -\frac{1}{3}\frac{2}{3} \right\} \qquad \text{and} \qquad G_1 := \left\{ x \in Y : f_1(x) \geq \frac{1}{3}\frac{2}{3} \right\}.$$

From Corollary 4.1.3 again, we obtain a continuous function $g_1 : X \to$ $[-\frac{1}{3}\frac{2}{3}, \frac{1}{3}\frac{2}{3}]$ with $g_1|_{F_1} = -\frac{1}{3}\frac{2}{3}$ and $g_1|_{G_1} = \frac{1}{3}\frac{2}{3}$. Set $f_2 := f_1 - g_1|_Y$ and observe that $f_2(Y) \subset [-\frac{2}{3}\frac{2}{3}, \frac{2}{3}\frac{2}{3}]$. Continuing in this fashion, we obtain continuous functions $f_0, f_1, f_2, \ldots$ on Y and $g_0, g_1, g_2, \ldots$ on X with

$$f_n(Y) \subset \left[-\left(\frac{2}{3}\right)^n, \left(\frac{2}{3}\right)^n \right] \qquad \text{and} \qquad g_n(X) \subset \left[-\frac{1}{3}\left(\frac{2}{3}\right)^n, \frac{1}{3}\left(\frac{2}{3}\right)^n \right]$$

for $n \in \mathbb{N}_0$. Moreover, we have

$$f_n = f_0 - (g_0 + g_1 + \cdots + g_{n-1})|_Y \qquad (n \in \mathbb{N}).$$

Let $\epsilon > 0$. Since

$$\sum_{n=0}^{\infty} \|g_n\|_\infty \le \sum_{n=0}^{\infty} \frac{1}{3}\left(\frac{2}{3}\right)^n = 1 < \infty, \qquad (*)$$

there is $n_\epsilon \in \mathbb{N}$ such that

$$\left\| \sum_{k=n+1}^{m} g_k \right\|_\infty \le \sum_{k=n+1}^{m} \|g_k\|_\infty < \epsilon \qquad (m > n \ge n_\epsilon).$$

Consequently, the sequence $\left(\sum_{k=0}^{n} g_k\right)_{n=1}^{\infty}$ is a Cauchy sequence in the Banach space $C_b(X, \mathbb{R})$ and therefore converges to a function $\tilde{f}: X \to \mathbb{R}$. By $(*)$, we have for $x \in X$ that

$$\left| \tilde{f}(x) \right| \le \sum_{n=0}^{\infty} |g_n(x)| \le \|g_n\|_\infty = 1,$$

so that $\tilde{f}(X) \subset [-1, 1]$. Moreover, for $x \in Y$,

$$\left| f(x) - \tilde{f}(x) \right| = \lim_{n\to\infty} \left| f(x) - \sum_{k=0}^{n} g_k(x) \right| = \lim_{n\to\infty} |f_n(x)| \le \lim_{n\to\infty} \left(\frac{2}{3}\right)^n = 0$$

holds. Hence, $\tilde{f}$ extends f as claimed. $\square$

The following is a nice consequence of Urysohn's lemma and Tietze's extension theorem and shows that normality is precisely the condition that makes these two results work.

Corollary 4.1.14. *The following are equivalent for a T_1-space $(X, \mathcal{T})$.*

(i) *X is normal.*
(ii) *For any closed and disjoint $F, G \subset X$, there is a continuous function $f: X \to \mathbb{R}$ such that $f(X) \subset [0, 1]$, $f|_F = 0$, and $f|_G = 1$.*
(iii) *For any closed and disjoint $F, G \subset X$, there is a continuous function $f: X \to \mathbb{R}$ such that $f|_F = 0$ and $f|_G = 1$.*

(iv) *For any closed subspace Y of X and for any continuous function $f: Y \to \mathbb{R}$ with $f(Y) \subset [a, b]$, there is a continuous function $\tilde{f}: X \to \mathbb{R}$ with $\tilde{f}(X) \subset [a, b]$ such that $\tilde{f}|_Y = f$.*

(v) *For any closed subspace Y of X and for any continuous function $f: Y \to \mathbb{R}$, there is a continuous function $\tilde{f}: X \to \mathbb{R}$ such that $\tilde{f}|_Y = f$.*

Proof. (i) $\implies$ (ii) is Uryhsohn's lemma, and (ii) $\implies$ (iii) is a triviality.

(iii) $\implies$ (i): Let $F, G \subset X$ be closed and disjoint, and let f be a function as in (iii). Set

$$U := \left\{ x \in X : f(x) < \frac{1}{2} \right\} \quad \text{and} \quad V := \left\{ x \in X : f(x) > \frac{1}{2} \right\}.$$

Then U and V are open and disjoint such that $F \subset U$ and $G \subset V$.

(i) $\implies$ (iv) is Tietze's extension theorem.

(iv) $\implies$ (v): Let $g := \arctan \circ f$. Then $g: Y \to \mathbb{R}$ is continuous such that $g(Y) \subset \left(-\frac{\pi}{2}, \frac{\pi}{2} \right)$. By (iv), there is a continuous function $\tilde{g}: X \to \mathbb{R}$ extending g such that $\tilde{g}(X) \subset \left[-\frac{\pi}{2}, \frac{\pi}{2} \right]$.

One might be tempted to simply set $\tilde{f} := \tan \circ \tilde{g}$. The problem with this approach is that, even though g does not attain $-\frac{\pi}{2}$ or $\frac{\pi}{2}$, we cannot rule out by (iv) that these two values do not lie in the range of $\tilde{g}$, so that $\tan \circ \tilde{g}$ may not be defined on all of X. To get around this difficulty, we invoke (iv) a second time.

Let $F := \tilde{g}^{-1} \left(\left\{ -\frac{\pi}{2}, \frac{\pi}{2} \right\} \right)$. Then F is closed and has an empty intersection with Y. Define $h: F \cup Y \to [0, 1]$ such that $h|_F = 0$ and $h|_Y = 1$. Then h is continuous because F and Y are clopen in $F \cup Y$, and (iv) yields a continuous extension $\tilde{h}: X \to [0, 1]$ of h. It follows that $\tilde{h}\tilde{g}$ is a continuous extension of g attaining all its values in $\left(-\frac{\pi}{2}, \frac{\pi}{2} \right)$. Consequently, $\tilde{f} := \tan \circ \left(\tilde{h}\tilde{g} \right)$ is a continuous extension of f.

(v) $\implies$ (iii): Let $F, G \subset X$ be closed and disjoint. Set $Y := F \cup G$ and define $f: Y \to \mathbb{R}$ through $f|_F := 0$ and $f|_G := 1$. Then f is continuous, and thus has a continuous extension to all of X. This extension satisfies (iii). $\square$

Exercises

1. Let $(X, \mathcal{T})$ be a T_1-space such that, for each closed $F \subset X$ and each open $U \subset X$ with $F \subset U$, there is an open subset V of X with $F \subset V \subset \overline{V} \subset U$. Show that X is normal.

2. Show that an open subset of a compact Hausdorff space is locally compact, and conclude that, for a locally compact Hausdorff space $(X, \mathcal{T})$, the neighborhood system $\mathcal{N}_x$ has a base consisting of compact sets for each $x \in X$.

3. Let (X, d) be a metric space (so that X is normal and Urysohn's lemma holds), and let $F, G \subset X$ be nonempty, closed, and disjoint. Give a "concrete" description of f as in Urysohn's lemma in terms of $\text{dist}(\cdot, F)$ and $\text{dist}(\cdot, G)$.

4. Let $(X, \mathcal{T})$ be a completely regular space, let $K \subset X$ be compact, and let $F \subset X$ be closed such that $K \cap F = \varnothing$. Show that there is a continuous function $f: X \to \mathbb{R}$ such that $f(X) \subset [0, 1]$, $f|_K = 1$, and $f|_F = 0$.

5. Let $(X, \mathcal{T})$ be a normal space, and let $\{U_1, \ldots, U_n\}$ be an open cover for X.
    (a) Show that there is an open cover $\{V_1, \ldots, V_n\}$ of X such that $\overline{V}_j \subset U_j$ for $j = 1, \ldots, n$.
    (b) Show that there are continuous functions $f_1, \ldots, f_n \colon X \to [0, \infty)$ such that $f_1 + \cdots + f_n = 1$ and $\overline{\{x \in X : f_j(x) \neq 0\}} \subset U_j$ for $j = 1, \ldots, n$.
6. Show that the following are equivalent for a topological space $(X, \mathcal{T})$.
    (i) X is completely regular.
    (ii) X is homeomorphic to a subspace of $[0, 1]^{\mathbb{I}}$ for some index set $\mathbb{I}$.
    (iii) X is homeomorphic to a subspace of a compact Hausdorff space.
7. Show that a Hausdorff space $(X, \mathcal{T})$ has a base for $\mathcal{T}$ consisting of clopen sets if and only if X is homeomorphic to a subspace of $\{0, 1\}^{\mathbb{I}}$ for some index set $\mathbb{I}$ (here, $\{0, 1\}$ is equipped with the discrete topology).
8. Give an example showing that, in Tietze's extension theorem, the requirement that Y be closed cannot be dropped.
9. Let $(X, \mathcal{T})$ be a normal space, and let Y be a closed subspace of X. Show that the restriction map

$$C_b(X, \mathbb{R}) \to C_b(Y, \mathbb{R}), \quad f \mapsto f|_Y$$

is continuous and surjective, and conclude that, if $C_b(X, \mathbb{R})$ is separable, then so is $C_b(Y, \mathbb{R})$. What if we replace $\mathbb{R}$ by $\mathbb{C}$?

4.2 The Stone–Čech Compactification

In Theorem 3.3.26, we saw that every locally compact Hausdorff space X is a subspace of a compact Hausdorff space, namely its one-point compactification X_∞. This compactification is minimal: it just contains one point not contained in X.

By Exercise 4.1.6, the topological spaces that have a "compactification" (i.e., are homeomorphic to a subspace of a compact Hausdorff space) are precisely the completely regular ones. In this section, we show that, among the compactifications of a completely regular space, one particular compactification stands out—in a sense yet to be made precise—as maximal.

We start with a look at the continuous functions on a compact Hausdorff space. This set of functions is a commutative ring with identity under the pointwise operations; we may thus speak of ideals in this ring.

Proposition 4.2.1. *Let $(K, \mathcal{T})$ be a compact Hausdorff space. Then the following are equivalent for $\mathfrak{m} \subset C(K, \mathbb{F})$.*

(i) $\mathfrak{m}$ *is a maximal ideal of* $C(K, \mathbb{F})$.
(ii) *There is a nonzero, multiplicative linear map* $\phi \colon C(K, \mathbb{F}) \to \mathbb{F}$ *such that* $\mathfrak{m} = \{f \in C(K, \mathbb{F}) : \phi(f) = 0\}$.
(iii) *There is $x \in K$ such that* $\mathfrak{m} = \{f \in C(K, \mathbb{F}) : f(x) = 0\}$.

Proof. (iii) $\Longrightarrow$ (ii): The map $C(K, \mathbb{F}) \ni f \mapsto f(x)$ is nonzero (because the constant functions are continuous), linear, and multiplicative, and has the required property.

(ii) $\Longrightarrow$ (i): It is routinely checked that $\mathfrak{m}$ is an ideal of $C(K, \mathbb{F})$. To see that $\mathfrak{m}$ is indeed maximal, first note that ϕ is surjective: ϕ is nonzero and linear, therefore the range of ϕ is a nonzero linear subspace of the one-dimensional linear space $\mathbb{F}$ and thus all of $\mathbb{F}$. Let I be an ideal of $C(K, \mathbb{F})$ with $\mathfrak{m} \subsetneqq I$. Since ϕ is surjective, $\phi(I)$ is an ideal of $\mathbb{F}$, and since $I \neq \mathfrak{m}$, it cannot be that $\phi(I)$ is the zero ideal. Since $\mathbb{F}$ is a field and thus does not have ideals besides (0) and $\mathbb{F}$, this means that $\phi(I) = \mathbb{F}$. Assume that $I \subsetneqq C(K, \mathbb{F})$, and let $f \in C(K, \mathbb{F}) \setminus I$. Since $\phi(I) = \mathbb{F}$, there is $g \in I$ such that $\phi(g) = \phi(f)$. This, in turn, implies that $f - g \in \mathfrak{m}$, so that

$$f = g + \underbrace{(f - g)}_{\in \mathfrak{m} \subset I} \in I,$$

which is a contradiction.

(i) $\Longrightarrow$ (iii): For each $x \in K$, let

$$\mathfrak{m}_x := \{ f \in C(K, \mathbb{F}) : f(x) = 0 \}.$$

Assume that $\mathfrak{m} \neq \mathfrak{m}_x$ for all $x \in K$. Since $\mathfrak{m}$ is maximal, this is equivalent to $\mathfrak{m} \not\subset \mathfrak{m}_x$ for each $x \in K$. Hence, for each $x \in K$, there is $f_x \in \mathfrak{m}$ with $f_x(x) \neq 0$. For $x \in K$, let

$$U_x := \{ y \in K : f_x(y) \neq 0 \}.$$

Then $\{ U_x : x \in K \}$ is an open cover for K and thus has a finite subcover: there are $x_1, \ldots, x_n \in K$ such that

$$K = U_{x_1} \cup \cdots \cup U_{x_n}.$$

Let

$$f := f_{x_1} \bar{f}_{x_1} + \cdots + f_{x_n} \bar{f}_{x_n}.$$

It follows that $f \in \mathfrak{m}$ and that $f(x) > 0$ for all $x \in K$. Since $\frac{1}{f}$ is again continuous, it follows that $1 = f \frac{1}{f} \in \mathfrak{m}$, which is impossible if $\mathfrak{m} \neq C(K, \mathbb{F})$. $\square$

Corollary 4.2.2. *Let $(K, \mathcal{T})$ be a compact Hausdorff space, and let $\phi : C(K, \mathbb{F}) \to \mathbb{F}$ be a nonzero, linear, and multiplicative map. Then there is a unique $x \in K$ such that*

$$\phi(f) = f(x) \qquad (f \in C(K, \mathbb{F})).$$

Proof. Let $\mathfrak{m} = \{ f \in C(K, \mathbb{F}) : \phi(f) = 0 \}$. By Proposition 4.2.1, there is $x \in K$ such that $\mathfrak{m} = \{ f \in C(K, \mathbb{F}) : f(x) = 0 \}$. Note that $\phi(1) = 1$: since $\phi(1)^2 = \phi(1^2) = \phi(1)$, it follows that $\phi(1) \in \{0, 1\}$, and $\phi(1) = 0$ is impossible because otherwise, $\phi(f) = \phi(f)\phi(1) = 0$ would hold for each $f \in C(K, \mathbb{F})$. Let $f \in C(K, \mathbb{F})$ be arbitrary. It follows that $f - f(x)1 \in \mathfrak{m}$, so that

$$\phi(f) = \phi(f - f(x)1) + \phi(f(x)1) = f(x)\phi(1) = f(x).$$

This proves the existence of x. The uniqueness follows from Urysohn's lemma. $\square$

The following proposition provides information on certain maps between spaces of continuous functions.

Proposition 4.2.3. *Let* $(K, \mathcal{T}_K)$ *and* $(L, \mathcal{T}_L)$ *be compact Hausdorff spaces. Then the following are equivalent for a map* $\phi \colon C(K, \mathbb{F}) \to C(L, \mathbb{F})$:

(i) *ϕ is linear, unital ring homomorphism, that is, is linear, multiplicative, and maps the identity of $C(K, \mathbb{F})$ to the identity of $C(L, \mathbb{F})$.*

(ii) *There is a continuous map* $\kappa \colon L \to K$ *such that*

$$\phi(f) = f \circ \kappa \qquad (f \in C(K, \mathbb{F})).$$

Moreover, κ as in (ii) is necessarily unique.

Proof. (ii) $\Longrightarrow$ (i) is trivial.

(i) $\Longrightarrow$ (ii): For any $x \in L$, define

$$\phi_x \colon C(K, \mathbb{F}) \to \mathbb{F}, \quad f \mapsto \phi(f)(x).$$

It is immediately checked, that ϕ_x is nonzero, linear, and multiplicative. By Corollary 4.2.2, there is therefore $\kappa(x) \in K$ with the property that

$$\phi(f)(x) = f(\kappa(x)) \qquad (f \in C(K, \mathbb{F})).$$

It remains to be shown that $\kappa \colon L \to K$ is continuous. What is clear is that κ is continuous if K is equipped with the coarsest topology making all functions in $C(K, \mathbb{F})$ continuous. This topology, however, is nothing but $\mathcal{T}_K$ by Corollary 4.1.8.

Finally, the uniqueness assertion of Corollary 4.2.2 yields the uniqueness of κ. $\square$

We can now formulate (and prove) the main result of this section.

Theorem 4.2.4 (Stone–Čech compactification). *Let* $(X, \mathcal{T}_X)$ *be a completely regular topological space. Then there is a compact Hausdorff space βX—the* Stone–Čech compactification *of X—along with a continuous map $\iota \colon X \to \beta X$, which is a homeomorphism onto a dense subset of βX, with the following universal property. If $(K, \mathcal{T}_K)$ is any compact Hausdorff space and if $\kappa \colon X \to K$ is continuous, then there is a unique continuous map $\hat{\kappa} \colon \beta X \to K$ such that the diagram*

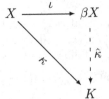

commutes. Moreover, βX is unique up to homeomorphism.

Proof. Let $\tilde{X} := \prod_{f \in C_b(X,\mathbb{R})} \overline{f(X)}$ be equipped with the product topology, so that $\tilde{X}$ is compact by Tychonoff's theorem, and define

$$\iota: X \to \tilde{X}, \quad x \mapsto (f(x))_{f \in C_b(X,\mathbb{R})}.$$

By the definition of the product topology, ι is continuous. Moreover, since X is completely regular, ι is injective, and Corollary 4.1.8 immediately yields that ι is a homeomorphism onto its image. Let $\beta X := \overline{\iota(X)}$. For $f \in C_b(X,\mathbb{R})$, let $\pi_f: \tilde{X} \to \mathbb{R}$ be the coordinate projection associated with f. Define $\hat{f}: \beta X \to \mathbb{R}$ by letting $\hat{f}(\omega) := \pi_f(\omega)$ for $\omega \in \beta X$. It follows that $\hat{f} \circ \iota = f$. Identifying X with its image in βX, we see that $\hat{f}$ is a (necessarily unique) continuous extension of f to all of βX. It is easy to see that

$$C_b(X,\mathbb{R}) \to C(\beta X,\mathbb{R}), \quad f \mapsto \hat{f}$$

is a linear, unital ring isomorphism.

Let $(K, \mathcal{T}_K)$ be any compact Hausdorff space, and let $\kappa: X \to K$ be continuous. Then

$$C(K,\mathbb{R}) \to C_b(X,\mathbb{R}), \quad f \mapsto f \circ \kappa$$

is a linear, unital ring homomorphism. In view of the isomorphism between $C_b(X,\mathbb{R})$ and $C(\beta X,\mathbb{R})$, Proposition 4.2.3 yields a continuous map $\hat{\kappa}: \beta X \to K$ such that

$$f(\hat{\kappa}(\iota(x))) = f(\kappa(x)) \qquad (f \in C(K,\mathbb{R}),\, x \in X).$$

It follows that $\hat{\kappa} \circ \iota = \kappa$.

Suppose that there are $\hat{\kappa}_1, \hat{\kappa}_2: \beta X \to K$ such that $\hat{\kappa}_j \circ \iota = \kappa$ for $j = 1, 2$. Then κ_1 and κ_2 coincide on the dense subset $\iota(X)$ and (because K is Hausdorff) must therefore be equal (Exercise 3.2.10). This proves the uniqueness of $\hat{\kappa}$.

To prove the uniqueness of βX up to homeomorphism, let $\beta_1 X$ and $\beta_2 X$ be Stone–Čech compactifications of X with corresponding maps $\iota_j: X \to \beta_j X$ for $j = 1, 2$. Then ι_1 and ι_2 have continuous extensions $\hat{\iota}_1: \beta_2 X \to \beta_1 X$ and $\hat{\iota}_2: \beta_1 X \to \beta_2 X$ such that $\hat{\iota}_1 \circ \iota_2 = \iota_1$ and $\hat{\iota}_2 \circ \iota_1 = \iota_2$. It follows that $\hat{\iota}_1$ and $\hat{\iota}_2$ are inverse to each other, so that $\beta_1 X$ and $\beta_2 X$ are homeomorphic via $\hat{\iota}_1$ and $\hat{\iota}_2$. $\square$

The following is a byproduct of the proof of Theorem 4.2.4, at least for $\mathbb{F} = \mathbb{R}$, but we rather deduce it from that theorem.

Corollary 4.2.5. *Let $(X, \mathcal{T})$ be a completely regular topological space. Then each $f \in C_b(X,\mathbb{F})$ has a unique extension $\hat{f} \in C(\beta X,\mathbb{F})$, so that*

$$C_b(X,\mathbb{F}) \to C(\beta X,\mathbb{F}), \quad f \mapsto \hat{f} \tag{$**$}$$

is a linear, unital ring isomorphism and an isometry of Banach spaces.

Proof. For $f \in C_b(X, \mathbb{F})$, apply Theorem 4.2.4 to $K := \overline{f(X)}$ and $\kappa := f$. The uniqueness of $\hat{f}$ follows from the denseness of X in βX, which also yields that $(**)$ is a linear, unital ring homomorphism and an isometry of Banach spaces. If $g \in C(\beta X, \mathbb{F})$, then its restriction f to X lies in $C_b(X, \mathbb{F})$, and the uniqueness of $\hat{f}$ yields that $\hat{f} = g$. Hence, $(**)$ is bijective. $\quad\square$

With the help of the Stone–Čech compactification, we can now tie up one loose end from the section on separation properties. With the exception of normality, it is easy to see that all the separation properties we discussed were inherited by arbitrary subspaces. For normality, this is not only not obvious, but false.

Example 4.2.6. Let $(X, \mathcal{T})$ be any completely regular space that fails to be normal, the Sorgenfrey plane from Example 3.5.14, for instance. By Theorem 4.2.4, we can identify X with a dense subspace of its Stone–Čech compactification βX. As a compact Hausdorff space, βX is normal whereas its subspace X isn't.

In general, the existence of the Stone–Čech compactification is a mixed blessing. When dealing with bounded continuous functions on a completely regular space, one can replace the given space by a compact Hausdorff space, a generally better-behaved object. On the other hand, for most completely regular spaces, the Stone–Čech compactification is so enormously large and removed from intuition that little can be said beyond that it exists and is compact. The exercises at the end of this section give an inkling of how huge and bizarre the space $\beta \mathbb{N}$ is.

Exercises

1. Let $(K, \mathcal{T})$ be a compact Hausdorff space. Then, by Proposition 4.2.1,

$$\mathfrak{m}_x := \{f \in C(K, \mathbb{F}) : f(x) = 0\}$$

is a maximal, and therefore prime, ideal of the commutative ring $C(K, \mathbb{F})$ for each $x \in K$. Show that

$$K \to \operatorname{Spec}(C(K, \mathbb{F})), \quad x \mapsto \mathfrak{m}_x$$

is a homeomorphism onto the subspace of $\operatorname{Spec}(C(K, \mathbb{F}))$ consisting of the maximal ideals of $C(K, \mathbb{F})$. (*Hint:* Show that the map is continuous and then that $\{\mathfrak{m}_x : x \in K\}$ equipped with the topology inherited from $\operatorname{Spec}(C(K, \mathbb{F}))$ is Hausdorff.)

2. Let $(K, \mathcal{T})$ be a separable compact Hausdorff space. Show that there is a continuous surjection from $\beta \mathbb{N}$ onto K, where $\mathbb{N}$ is equipped with the discrete topology.

3. An *idempotent* in a commutative ring R with identity is an element $e \in R$ such that $e^2 = e$; for example, 0 and 1 are idempotents. Show that a topological space $(X, \mathcal{T})$ is connected if and only if $C_b(X, \mathbb{F})$ has no idempotents other than the constant functions 0 and 1, and conclude that a completely regular space is connected if and only if its Stone–Čech compactification is.

4. Let $(X, \mathcal{T})$ be a discrete topological space.
   (a) Show that the linear span of the idempotents in $C_b(X, \mathbb{F})$ is dense in $C_b(X, \mathbb{F})$.
   (b) Conclude that βX is zero-dimensional in the sense of Exercise 3.4.10 (and thus totally disconnected).
5. Let $\mathbb{N}$ be equipped with the discrete topology.
   (a) Show that a *sequence* in $\mathbb{N}$ converges in $\beta \mathbb{N}$ if and only if it is eventually constant.
   (b) Let $x \in \beta \mathbb{N} \setminus \mathbb{N}$. Show that $\mathcal{N}_x$ does not have a countable base (so that $\beta \mathbb{N}$ is not first countable and thus not metrizable).
6. Show that $\beta \mathbb{N}$ has the same cardinality as $\mathbb{R}^{\mathbb{R}}$. Proceed as follows.
   (a) Show that $[0, 1]^{[0,1]}$, equipped with the product topology, is a separable, compact Hausdorff space, and conclude that $|\beta \mathbb{N}| \geq |[0,1]^{[0,1]}| = |\mathbb{R}^{\mathbb{R}}|$.
   (b) Show that $C_b(\mathbb{N}, \mathbb{R}) = B(\mathbb{N}, \mathbb{R})$ has cardinality $\mathfrak{c}$, and conclude that also $|\beta \mathbb{N}| \leq |\mathbb{R}^{\mathbb{R}}|$ (so that, in fact, $|\beta \mathbb{N}| = |\mathbb{R}^{\mathbb{R}}|$ by the Cantor–Bernstein theorem).

4.3 The Stone–Weierstraß Theorems

In Example 2.4.18, we used the classical Weierstraß approximation theorem: Every continuous function on a compact interval can be uniformly approximated by a sequence of polynomials.

In this section, we extend this result to continuous functions on arbitrary compact Hausdorff spaces. The first question that arises here is, of course, what is a polynomial supposed to be on an arbitrary compact Hausdorff space.

Definition 4.3.1. *An algebra is a commutative ring A that is also a vector space over $\mathbb{F} = \mathbb{R}$ or $\mathbb{F} = \mathbb{C}$ such that*

$$\lambda(ab) = (\lambda a)b = a(\lambda b) \qquad (\lambda \in \mathbb{F}, \, a, b \in A).$$

If A has an identity, we call A unital. A linear subspace B of A that is also a subring, is called a subalgebra; if A has an identity and if this identity is contained in B, we call B a unital subalgebra.

Of course, the continuous functions on a compact interval form a unital algebra, the polynomials form a unital subalgebra, and the Weierstraß approximation theorem asserts nothing but that the subalgebra of the continuous functions consisting of the polynomials is dense. It is in this direction that we generalize the Weierstraß approximation theorem.

We start with a definition.

Definition 4.3.2. *Let $(K, \mathcal{T})$ be a compact Hausdorff space, and let A be a closed unital subalgebra of $C(K, \mathbb{C})$. Then $\varnothing \neq S \subset K$ is called A-antisymmetric if $f \in A$ with $f(S) \subset \mathbb{R}$ means that f is constant on S.*

Of course, no matter what A is, the singleton subsets of K are trivially A-antisymmetric. Our first lemma ascertains the existence of "large" A-antisymmetric sets for any closed subalgebra A of $C(K, \mathbb{C})$. To make precise what we mean by "large", we introduce some notation.

Let $(K, \mathcal{T})$ be any compact Hausdorff space, let $\varnothing \neq F \subset K$ be closed, and define

$$\|f\|_F = \sup\{|f(x)| : x \in F\} \qquad (f \in C(K, \mathbb{C})).$$

Clearly, if $F = K$, then $\|\cdot\|_K$ is just the norm $\|\cdot\|_\infty$. For $F \subsetneq K$, there are nonzero $f \in C(K, \mathbb{C})$ such that $\|f\|_F = 0$ by Uhrysohn's lemma. Hence, $\|\cdot\|_F$ is generally not a norm, but only what is called a *seminorm*; nevertheless, it still satisfies the triangle inequality and scalars come out as their absolute values. Let E be a subspace of $C(K, \mathbb{C})$, and let $f \in C(K, \mathbb{C})$. We define

$$\operatorname{dist}_F(f, E) := \inf\{\|f - g\|_F : g \in E\}.$$

Of course, if $F = K$, then $\operatorname{dist}_F(f, E)$ is nothing but the distance $\operatorname{dist}(f, E)$.

Lemma 4.3.3. *Let $(K, \mathcal{T})$ be a compact Hausdorff space, let $f \in C(K, \mathbb{C})$, and let A be a closed unital subalgebra of $C(K, \mathbb{C})$ (over $\mathbb{C}$). Then there is a closed, A-antisymmetric subset F of K such that $\operatorname{dist}_F(f, A) = \operatorname{dist}(f, A)$.*

Proof. Let

$$\mathcal{F} := \{\varnothing \neq F \subset K : F \text{ is closed such that } \operatorname{dist}_F(f, A) = \operatorname{dist}(f, A)\}.$$

Trivially, $K \in \mathcal{F}$, so that $\mathcal{F} \neq \varnothing$. Let $\mathcal{F}$ be ordered by reversed set inclusion; that is,

$$F_1 \preceq F_2 \quad :\Longleftrightarrow \quad F_2 \subset F_1 \qquad (F_1, F_2 \in \mathcal{F}).$$

Let $\mathcal{G}$ be a totally ordered subset of $\mathcal{F}$, and set $G_0 := \bigcap\{G : G \in \mathcal{G}\}$. It is clear that G_0 is again closed, and since K has the finite intersection property and $\mathcal{G}$ is totally ordered, it follows that $G_0 \neq \varnothing$. We claim that $G_0 \in \mathcal{F}$. To see this, fix $g \in A$, and note that, for any $G \in \mathcal{G}$, the set

$$\{x \in G : |f(x) - g(x)| \geq \operatorname{dist}(f, A)\}$$

is compact and not empty. Again invoking the finite intersection property of K, we obtain that

$$\{x \in G_0 : |f(x) - g(x)| \geq \operatorname{dist}(f, A)\} = \bigcap_{G \in \mathcal{G}} \{x \in G : |f(x) - g(x)| \geq \operatorname{dist}(f, A)\}$$

is also compact and nonempty, so that $\|f - g\|_{G_0} \geq \operatorname{dist}(f, A)$. Since $g \in A$ was arbitrary, we conclude that $\operatorname{dist}_{G_0}(f, A) \geq \operatorname{dist}(f, A)$. Since the reversed inequality holds trivially, we see that $G_0 \in \mathcal{F}$. By Zorn's lemma, $\mathcal{F}$ thus has maximal (meaning: minimal with respect to set inclusion) elements. Let $F \in \mathcal{F}$ be such a minimal set.

We claim that F is A-antisymmetric. Assume otherwise, so that there is a function $g \in A$ which is real-valued, but not constant on F. Without loss of generality, suppose that $g(F) \subset [0,1]$, and that $[0,1]$ is the smallest interval containing $g(F)$. Set

$$F_1 := \left\{ x \in F : g(x) \leq \frac{2}{3} \right\} \qquad \text{and} \qquad F_2 := \left\{ x \in F : g(x) \geq \frac{1}{3} \right\}.$$

Then F_1 and F_2 are nonempty, proper closed subsets of F whose union is all of F. By the minimality of F, there are $h_1, h_2 \in A$ with $\|f - h_j\|_{F_j} < \mathrm{dist}(f, A)$ for $j = 1, 2$. For $n \in \mathbb{N}$, define

$$g_n := (1 - g^n)^{2^n} \qquad \text{and} \qquad k_n := g_n h_1 + (1 - g_n) h_2,$$

so that $g_n, k_n \in A$ for all $n \in \mathbb{N}$. We show that $\|f - k_n\|_F < \mathrm{dist}(f, A)$ if $n \in \mathbb{N}$ is sufficiently large and thus arrive at a contradiction.

First, consider $x \in F_1 \cap F_2$ and note that

$$
\begin{aligned}
&|f(x) - k_n(x)| \\
&= |g_n(x)f(x) + (1 - g_n(x))f(x) - g_n(x)h_1(x) + (1 - g_n(x))h_2(x)| \\
&\leq g_n(x)|f(x) - h_1(x)| + (1 - g_n(x))|f(x) - h_2(x)| \\
&\leq g_n(x)\|f - h_1\|_{F_1} + (1 - g_n(x))\|f - h_2\|_{F_2} \\
&< \mathrm{dist}(f, A)
\end{aligned}
$$

for all $n \in \mathbb{N}$.

For $x \in F_1 \setminus F_2$, we have

$$
\begin{aligned}
1 \geq g_n(x) \\
\geq 1 - 2^n g(x)^n, \qquad \text{by Bernoulli's inequality,} \\
\geq 1 - \left(\frac{2}{3} \right)^n \qquad (n \in \mathbb{N}).
\end{aligned}
$$

It follows that $\|g_n - 1\|_{F_1 \setminus F_2} \to 0$ and thus $\|k_n - h_1\|_{F_1 \setminus F_2} \to 0$. Since $\|f - h_1\|_{F_1} < \mathrm{dist}(f, A)$, we obtain $n_1 \in \mathbb{N}$ such that $\|f - k_n\|_{F_1 \setminus F_2} < \mathrm{dist}(f, A)$ for all $n \geq n_1$.

For $x \in F_2 \setminus F_1$, first observe that $g_n(x) \leq \frac{1}{(1+g(x)^n)^{2^n}}$ because

$$g_n(x)(1 + g(x)^n)^{2^n} = (1 - g(x)^n)^{2^n}(1 + g_n(x))^{2^n} = (1 - g(x)^{2n})^{2^n} \leq 1,$$

so that

$$0 \le g_n(x)$$

$$\le \frac{1}{(1 + g(x)^n)^{2^n}}$$

$$\le \frac{1}{1 + 2^n g(x)^n}, \qquad \text{again by Bernoulli's inequality,}$$

$$\le \frac{1}{2^n g(x)^n}$$

$$\le \left(\frac{3}{4}\right)^n \qquad (n \in \mathbb{N}).$$

Consequently, $\|g_n\|_{F_2 \setminus F_1} \to 0$ holds, so that $\|k_n - h_2\|_{F_2 \setminus F_1} \to 0$. Since $\|f - h_2\|_{F_2} < \operatorname{dist}(f, A)$, there is thus $n_2 \in \mathbb{N}$ such that $\|f - k_n\|_{F_2 \setminus F_1} < \operatorname{dist}(f, A)$ for all $n \ge n_2$.

Let $n \ge \max\{n_1, n_2\}$, and let $x_0 \in F$ be such that $|f(x_0) - k_n(x_0)| = \|f - k_n\|_F$. Since $F = F_1 \cup F_2$, there are three possibilities: $x_0 \in F_1 \cap F_2$, $x_0 \in F_1 \setminus F_2$, or $x_0 \in F_2 \setminus F_1$. By the foregoing estimates, we have in any of those cases that

$$\operatorname{dist}_F(f, A) \le \|f - k_n\|_F = |f(x_0) - k_n(x_0)| < \operatorname{dist}(f, A),$$

which is impossible if $F \in \mathcal{F}$. □

With Lemma 4.3.3 proven, the main result of this section is surprisingly easy to obtain.

Theorem 4.3.4 (complex Stone–Weierstraß theorem). *Let $(K, \mathcal{T})$ be a compact Hausdorff space, and let A be a subalgebra of $C(K, \mathbb{C})$ with the following properties.*

(a) *$1 \in A$,*
(b) *For any $x, y \in K$ with $x \neq y$, there is $f \in A$ such that $f(x) \neq f(y)$,*
(c) *If $f \in A$, then $\bar{f} \in A$, where $\bar{f}$ stands for pointwise conjugation.*

Then A is dense in $C(K, \mathbb{C})$.

Proof. Replace A by its closure; it is clear that this does not affect properties (a), (b), and (c).

Let $f \in C(K, \mathbb{C})$. By Lemma 4.3.3, there is an A-antisymmetric subset $\varnothing \neq F \subset K$ such that $\operatorname{dist}_F(f, A) = \operatorname{dist}(f, A)$. Assume that there are $x, y \in F$ with $x \neq y$. By (b), there is $f \in A$ such that $f(x) \neq f(y)$. Since $\bar{f} \in A$ by (c), both $\operatorname{Re} f = \frac{1}{2}(f + \bar{f})$ and $\operatorname{Im} f = \frac{1}{2i}(f - \bar{f})$ belong to A, and since $f = \operatorname{Re} f + i \operatorname{Im} f$, it is clear that $\operatorname{Re} f(x) \neq \operatorname{Re} f(y)$ or $\operatorname{Im} f(x) \neq \operatorname{Im} f(y)$. This is impossible, however, due to the definition of an A-antisymmetric set.

We conclude that F is a singleton set, say $\{x_0\}$. Since the constant function $f(x_0)1$ belongs to A, we see that

$$\operatorname{dist}(f, A) = \operatorname{dist}_F(f, A) \le \|f - f(x_0)1\|_F = 0,$$

so that $f \in A$. □

The demand that A be closed under pointwise conjugation cannot be dropped.

Example 4.3.5. Let $K = \{z \in \mathbb{C} : |z| \le 1\}$ (i.e., the closed unit disc in $\mathbb{C}$), and let A be the subalgebra of $C(K, \mathbb{C})$ consisting of those functions that are holomorphic on $\{z \in \mathbb{C} : |z| < 1\}$. This is a closed subalgebra of $C(K, \mathbb{C})$ that satisfies conditions (a) and (b) of Theorem 4.3.4, but nevertheless is not all of $C(K, \mathbb{C})$.

A real Stone–Weierstraß theorem is easily deduced from the complex one.

Corollary 4.3.6 (real Stone–Weierstraß theorem). *Let $(K, \mathcal{T})$ be a compact Hausdorff space, and let A be a subalgebra of $C(K, \mathbb{R})$ with the following properties.*

(a) $1 \in A$,
(b) *For any $x, y \in K$ with $x \ne y$, there is $f \in A$ such that $f(x) \ne f(y)$.*

Then A is dense in $C(K, \mathbb{R})$.

Proof. Let $A_{\mathbb{C}} := \{f + ig : f, g \in A\}$. Then $A_{\mathbb{C}}$ is a subalgebra of $C(K, \mathbb{C})$ that satisfies the requirements of the complex Stone–Weierstraß theorem and thus is dense in $C(K, \mathbb{C})$. This, in turn, yields that A is dense in $C(K, \mathbb{R})$. □

To appreciate the generality of Theorem 4.3.4 and Corollary 4.3.6, consider the following.

Corollary 4.3.7. *Let $K \subset \mathbb{R}^n$ be compact, and let $f \colon K \to \mathbb{F}$ be continuous. Then there is a sequence of polynomials in n variables and with coefficients in $\mathbb{F}$ that converges to f uniformly on K.*

Proof. Apply Theorem 4.3.4 or Corollary 4.3.6 to the algebra of all polynomials in n variables with coefficients in $\mathbb{F}$. □

And just for the record, we have the following.

Corollary 4.3.8 (Weierstraß approximation theorem). *Let $a, b \in \mathbb{R}$ with $a < b$, and let $f \colon [a, b] \to \mathbb{F}$ be continuous. Then there is a sequence of polynomials with coefficients in $\mathbb{F}$ that converges to f uniformly on $[a, b]$.*

Even though our approach to the Weierstraß approximation theorem (via Lemma 4.3.3 and the Stone–Weierstraß theorem) is very short and elegant (if not slick), it has its drawbacks: given a continuous function on a compact interval, we know that there somehow is a sequence of polynomials converging uniformly to f, but we have no information whatsoever on what such polynomials might look like for concrete f (Lemma 4.3.3 relies on Zorn's lemma). An alternative proof of the approximation theorem, which is both elementary, nothing beyond first-year calculus is required, and constructive, is outlined in Exercise 1 below.

As another application of the Stone–Weierstraß theorem, we present another characterization of metrizability for compact Hausdorff spaces.

Proposition 4.3.9. *The following are equivalent for a compact Hausdorff space* $(K, \mathcal{T})$.

(i) $C(K, \mathbb{F})$ *is separable.*
(ii) K *is metrizable.*

Proof. (i) $\implies$ (ii): Let $\{f_n : n \in \mathbb{N}\}$ be a countable dense subset of $C(K, \mathbb{F})$, and define

$$d: K \times K \to [0, \infty), \quad (x, y) \mapsto \sum_{n=1}^{\infty} \frac{1}{2^n} \frac{|f_n(x) - f_n(y)|}{1 + |f_n(x) - f_n(y)|}.$$

Then d is a metric on K such that $\mathrm{id}: (K, \mathcal{T}) \to (K, d)$ is continuous (and, trivially, bijective), so that, by Theorem 3.3.11, it is a homeomorphism.

(ii) $\implies$ (i): Suppose that K is metrizable. From Corollary 4.1.11, it follows that K is second countable. Let $\{U_1, U_2, \ldots\}$ be a countable base for $\mathcal{T}$, let (as in the proof of Urysohn's metrization theorem)

$$\mathbb{A} := \left\{ (n, m) \in \mathbb{N}^2 : \overline{U}_n \subset U_m \right\},$$

and (again as in the proof of the metrization theorem) choose, for each $(n, m) \in \mathbb{A}$, a continuous function $f_{n,m} : X \to \mathbb{R}$ with $f_{n,m}(X) \subset [0, 1]$, $f_{n,m}|_{\overline{U}_n} = 1$, and $f_{n,m}|_{X \setminus U_m} = 0$. Certainly, $S := \{f_{n,m} : (n, m) \in \mathbb{A}\}$ is countable, as is Π_S, the collection of all finite products of elements of S. If $\mathbb{F} = \mathbb{R}$, let $\mathbb{F}_0 := \mathbb{Q}$, and if $\mathbb{F} = \mathbb{C}$, let $\mathbb{F}_0 := \{q + ir : q, r \in \mathbb{Q}\}$; in either case, $\mathbb{F}_0$ is a countable subfield of $\mathbb{F}$. The linear combinations of elements from Π_S over $\mathbb{F}_0$ then form a countable subset of $C(K, \mathbb{F})$ whose closure is a— necessarily separable—subalgebra A of $C(K, \mathbb{F})$. It is immediate that $1 \in A$ and that, if $f \in A$, then $\bar{f} \in A$ (vacuous if $\mathbb{F} = \mathbb{R}$). Let $x, y \in K$ be such that $x \neq y$. As in the proof of Urysohn's metrization theorem, we see that there is $(n, m) \in \mathbb{A}$ such that $f_{n,m}(x) \neq f_{n,m}(y)$. By the complex and real Stone–Weierstraß theorems, A therefore equals $C(K, \mathbb{F})$, so that $C(K, \mathbb{F})$ is separable. $\square$

Concluding this section, we turn to locally compact spaces.

Definition 4.3.10. *Let* $(X, \mathcal{T})$ *be a locally compact Hausdorff space. A continuous function* $f : X \to \mathbb{F}$ *is said to* vanish at infinity *if, for each* $\epsilon > 0$, *there is a compact subset* K *of* X *such that* $|f(x)| < \epsilon$ *for all* $x \in X \setminus K$. *The collection of all continuous functions from* X *to* $\mathbb{F}$ *that vanish at infinity is denoted by* $C_0(X, \mathbb{F})$.

Examples 4.3.11. (a) Every continuous function on a compact space vanishes at infinity.
(b) A function f on the real line vanishes at infinity (in the sense of Definition 4.3.10) if and only if

$$\lim_{t \to \infty} f(t) = 0 = \lim_{t \to -\infty} f(t).$$

The following proposition explains the choice of terminology.

Proposition 4.3.12. *Let* $(X, \mathcal{T})$ *be a locally compact Hausdorff space with one-point compactification* X_∞. *Then a continuous function* $f\colon X \to \mathbb{F}$ *vanishes at infinity if and only if* f *has a continuous extension* $\tilde{f}\colon X_\infty \to \mathbb{F}$ *such that* $\tilde{f}(\infty) = 0$.

Proof. Suppose that f vanishes at infinity, and let $\tilde{f}$ be the unique extension of f to X_∞ such that $\tilde{f}(\infty) = 0$. We need to show that $\tilde{f}$ is continuous at ∞. Let $\epsilon > 0$ and choose $K \subset X$ compact such that $|f(x)| < \epsilon$ for all $x \in X \setminus K$. From the definition of the topology on X_∞, it follows that $X_\infty \setminus K$ belongs to $\mathcal{N}_\infty$. Consequently, $\tilde{f}^{-1}(B_\epsilon(0))$ is a neighborhood of ∞ for each $\epsilon > 0$.

Conversely, suppose that f has a continuous extension $\tilde{f}$ vanishing at ∞. Let $\epsilon > 0$. Then $\tilde{f}^{-1}(B_\epsilon(0))$ is an open neighborhood of ∞ and thus is of the form $X_\infty \setminus K$ for some compact set $K \subset X$; that is, $|f(x)| < \epsilon$ for $x \in X \setminus K$. □

We can thus identify the continuous functions vanishing at infinity on a locally compact Hausdorff space with those functions on its one-point compactification that vanish at the point ∞.

With this in mind, we can prove a version of the Stone–Weierstraß theorem for locally compact spaces, which we do simultaneously over both $\mathbb{R}$ and $\mathbb{C}$.

Theorem 4.3.13. *Let* $(X, \mathcal{T})$ *be a locally compact Hausdorff space, and let* A *be a subalgebra of* $C_0(X, \mathbb{F})$ *with the following properties.*

(a) *For any* $x \in X$, *there is* $f \in A$ *with* $f(x) \neq 0$,
(b) *For any* $x, y \in X$ *with* $x \neq y$, *there is* $f \in A$ *such that* $f(x) \neq f(y)$,
(c) *If* $f \in A$, *then* $\bar{f} \in A$, *where* $\bar{f}$ *stands for pointwise conjugation (which is vacuous if* $\mathbb{F} = \mathbb{R}$).

Then A *is dense in* $C_0(K, \mathbb{F})$.

Proof. In view of Proposition 4.3.12, we may identify $C_0(X, \mathbb{F})$ and thus A with a subalgebra of $C(X_\infty, \mathbb{F})$. Let $A^{\#} := \{f + \lambda 1 : f \in A, \lambda \in \mathbb{F}\}$, so that $A^{\#}$ is a subalgebra of $C(X_\infty, \mathbb{F})$ containing 1 and closed under pointwise conjugation. Let $x, y \in X_\infty$ with $x \neq y$. If $x, y \in X$, it follows from (b) that there is $f \in A \subset A^{\#}$ such that $f(x) \neq f(y)$. If $y = \infty$, (a) yields $f \in A \subset A^{\#}$ such that $f(x) \neq 0 = f(y)$. All in all, it follows from Theorem 4.3.4 and Corollary 4.3.6, respectively, that $A^{\#}$ is dense in $C(X_\infty, \mathbb{F})$. Let $f \in C_0(X, \mathbb{F})$ and let $\epsilon > 0$. Then there is $g \in A^{\#}$ such that $\|f - g\|_\infty < \frac{\epsilon}{2}$. Let $h := g - g(\infty)1$, so that $h \in A$. Since $f(\infty) = 0$, we have

$$|g(\infty)| \leq |f(\infty) - g(\infty)| \leq \|f - g\|_\infty < \frac{\epsilon}{2}.$$

We thus obtain that

$$\|f - h\|_\infty \leq \|f - g\|_\infty + |g(\infty)| < \frac{\epsilon}{2} + \frac{\epsilon}{2} = \epsilon.$$

Hence, A is dense in $C_0(X, \mathbb{F})$ as claimed. □

Since every compact space is locally compact, it follows that Theorem 4.3.13 also applies to compact spaces and improves Theorem 4.3.4 and Corollary 4.3.6, respectively: condition (i) of Theorem 4.3.13 is considerably weaker than condition (i) of both Theorem 4.3.4 and Corollary 4.3.6.

Exercises

1. *A constructive proof of the Weierstraß approximation theorem.* Let $f : [0, 1] \to \mathbb{R}$ be continuous. For $n \in \mathbb{N}_0$, the nth *Bernstein polynomial* of f is defined as

$$B_n(t) := \sum_{k=0}^{n} \binom{n}{k} t^k (1-t)^{n-k} f\left(\frac{k}{n}\right) \qquad (t \in [0, 1]).$$

Show that the sequence $(B_n)_{n=1}^{\infty}$ converges to f uniformly on $[0, 1]$. Proceed as follows.

(a) Successively prove the following identities for $n \in \mathbb{N}$ and $t \in [0, 1]$.

$$1 = \sum_{k=0}^{n} \binom{n}{k} t^k (1-t)^{n-k},$$

$$0 = \sum_{k=0}^{n} \binom{n}{k} t^k (1-t)^{n-k} (k - nt),$$

$$n = \sum_{k=0}^{n} \binom{n}{k} t^{k-1} (1-t)^{n-k-1} (k - nt)^2,$$

and

$$\frac{t(1-t)}{n} = \sum_{k=0}^{n} \binom{n}{k} t^k (1-t)^{n-k} \left(t - \frac{k}{n}\right)^2.$$

(*Hint*: Obtain the second identity through differentiation of the first one, and differentiate the second identity, to obtain the third one.)

(b) Show that

$$|B_n(t) - f(t)| \leq \sum_{k=0}^{n} \binom{n}{k} t^k (1-t)^{n-k} \left| f\left(\frac{k}{n}\right) - f(t) \right| \qquad (t \in [0, 1]).$$

(c) Fix $\epsilon > 0$, and (using the uniform continuity of f on $[0, 1]$) choose $\delta > 0$ such that $|f(s) - f(t)| < \frac{\epsilon}{2}$ for $s, t \in [0, 1]$ with $|s - t| < \delta$. Fix $t \in [0, 1]$, and let $N_\delta := \left\{ k \in \{0, \ldots, n\} : \left| \frac{k}{n} - t \right| < \delta \right\}$. Show that

$$\sum_{k \in N_\delta} \binom{n}{k} t^k (1-t)^{n-k} \left| f\left(\frac{k}{n}\right) - f(t) \right| < \frac{\epsilon}{2}.$$

(d) With ϵ, δ, t, and N_δ as in (c), show that

$$\sum_{k \notin N_\delta} \binom{n}{k} t^k (1-t)^{n-k} \left| f\left(\frac{k}{n}\right) - f(t) \right| \leq 2\|f\|_\infty \sum_{k \notin N_\delta} \binom{n}{k} t^k (1-t)^{n-k}$$

$$\leq \frac{2\|f\|_\infty}{\delta^2} \frac{t(1-t)}{n}$$

$$\leq \frac{2\|f\|_\infty}{4\delta^2 n}.$$

(e) Conclude that there is $n_\epsilon \in \mathbb{N}$ such that $|B_n(t) - f(t)| < \epsilon$ for all $t \in [0, 1]$ and $n \geq n_\epsilon$.

2. Let $(K, \mathcal{T})$ be a compact Hausdorff space. Show that K is zero-dimensional (see Exercise 3.4.10 for the definition) if and only if the linear span of the idempotents is dense in $C(K, \mathbb{F})$.

3. A *complex trigonometric polynomial* on $\mathbb{R}$ is a $\mathbb{C}$-valued function of the form

$$\sum_{k=-n}^{n} c_k e^{ikt} \qquad (t \in \mathbb{R}),$$

where $n \in \mathbb{N}$ and $c_{-n}, \ldots, c_n \in \mathbb{C}$. Use the complex Stone–Weierstraß theorem to show that, for every continuous 2π-periodic function $f : \mathbb{R} \to \mathbb{C}$, there is a sequence of complex trigonometric polynomials converging to f uniformly on $\mathbb{R}$.

4. A *real trigonometric polynomial* on $\mathbb{R}$ is an $\mathbb{R}$-valued function of the form

$$\sum_{k=-n}^{n} a_k \cos(kt) + b_k \sin(kt) \qquad (t \in \mathbb{R}),$$

where $n \in \mathbb{N}$ and $a_{-n}, b_{-n}, \ldots, a_n, b_n \in \mathbb{R}$. Show that, for every continuous 2π-periodic function $f : \mathbb{R} \to \mathbb{R}$, there is a sequence of real trigonometric polynomials converging to f uniformly on $\mathbb{R}$. (*Warning*: The space of all real trigonometric polynomials on $\mathbb{R}$ is *not* an algebra.)

5. Let $(X, \mathcal{T})$ be a locally compact Hausdorff space. Show that X is compact if and only if the constant function 1 belongs to $C_0(X, \mathbb{F})$.

6. Let $(X, \mathcal{T})$ be a locally compact Hausdorff space. Show that X is σ-compact (see Exercise 3.5.5) if and only if there is $f \in C_0(X, \mathbb{R})$ with $0 < f(x) \leq 1$ for $x \in X$.

7. Let $(X, \mathcal{T})$ be a locally compact Hausdorff space, and let $C_{00}(X, \mathbb{F})$ denote those continuous functions $f : X \to \mathbb{F}$ such that $f|_{X \setminus K} = 0$ for some compact set $K \subset X$.

   (a) Show that $C_0(X, \mathbb{F})$ and $C_{00}(X, \mathbb{F})$ are ideals in $C_b(X, \mathbb{F})$.

   (b) Show that $C_0(X, \mathbb{F})$ is closed in $C_b(X, \mathbb{F})$.

   (c) Show that $C_{00}(X, \mathbb{F})$ is dense in $C_0(X, \mathbb{F})$.

8. Let $(X, \mathcal{T}_X)$ and $(Y, \mathcal{T}_Y)$ be locally compact Hausdorff spaces, and let $X \times Y$ be equipped with the product topology. Show that $X \times Y$ is locally compact and that, for any $f \in C_0(X \times Y, \mathbb{F})$ and $\epsilon > 0$, there are $g_1, \ldots, g_n \in C_{00}(X, \mathbb{F})$ and $h_1, \ldots, h_n \in C_{00}(Y, \mathbb{F})$ with

$$\left| f(x, y) - \sum_{j=1}^{n} g_j(x) h_j(y) \right| < \epsilon \qquad (x \in X, \ y \in Y).$$

Remarks

Pavel S. Urysohn, after whom the "lemma" and the metrization theorem are named, was a close friend and collaborator of Alexandroff. Two years younger than Alexandroff, he was outlived by him by almost six decades; on a visit to France in 1924, Urysohn drowned while swimming in the sea. He was only 26.

Urysohn's metrization theorem gives a sufficient condition for metrizability, which is also necessary if the space is compact. A condition for the metrizability of general topological spaces that is both necessary and sufficient is given by the *Nagata–Smirnoff (Smirnov) theorem*, sometimes also called the *Bing–Nagata–Smirnoff theorem* (see [MUNKRES 00], for example).

The Stone–Čech compactification was discovered, independently, by Marshall H. Stone [STONE 37], an American, and Eduard Čech, Czech not only by name, but by ethnicity as well.

The monumental paper [STONE 37] also contains the generalization of the Weierstraß approximation theorem that would later become known as the Stone–Weierstraß theorem. Lemma 4.3.3, which lies at the heart of our proof, is due to the Brazilian mathematician Silvio Machado [MACHADO 77]; it was given the short and elegant proof we present by Thomas J. Ransford [RANSFORD 84].

Let $(K, \mathcal{T})$ be a compact Hausdorff space. By Exercise 4.2.1, the space K can be identified with the maximal ideals of the unital ring $C(K, \mathbb{C})$; that is, all information on K is already encoded in the algebraic structure of $C(K, \mathbb{C})$. This allows for a far-reaching generalization of topology, namely *noncommutative topology*.

An *involution* on an algebra A over $\mathbb{C}$ is a map

$$A \to A, \quad a \mapsto a^*$$

such that $(\lambda a + \mu b)^* = \bar{\lambda} a^* + \bar{\mu} b^*$, $(ab)^* = b^* a^*$, and $a^{**} = a$ for all $a, b \in A$ and $\lambda, \mu \in \mathbb{C}$. For example, if $A = C(K, \mathbb{C})$, then pointwise conjugation is an involution. Another example of an algebra with involution consists, for given $n \in \mathbb{N}$, of all $n \times n$ matrices over $\mathbb{C}$ (with entrywise addition and scalar multiplication, and with matrix multiplication as product): for any such matrix a, the matrix a^* is obtained by transposing a and conjugating its entries.

Suppose now that A is not only equipped with an involution, but also with a norm $\| \cdot \|$ that turns A into a Banach space and satisfies $\|ab\| \le \|a\|\|b\|$ and $\|a^* a\| = \|a\|^2$ for all $a, b \in A$. Then A is called a C^*-*algebra*. Of course, $C(K, \mathbb{C})$ is a commutative C^*-algebra. The surprising statement of the *Gelfand–Naimark theorem* is that *all* commutative unital C^*-algebras are of this form! Hence, a unital commutative C^*-algebra is nothing but a compact Hausdorff space in disguise: every statement about such an algebra translates into a statement about the associated space. For example, $A = C(K, \mathbb{C})$ has no nontrivial idempotents if and only if K is disconnected, and the linear span of the idempotents is dense in A if and only if K is zero-dimensional (Exercises 4.2.3 and 4.3.2).

Instead of considering commutative C^*-algebras as disguised topological spaces, one can take the opposite point of view and say that compact Hausdorff spaces are nothing but disguised C^*-algebras. This may seem artificial, but it opens up a whole new world of mathematical objects: the noncommutative C^*-algebras. Here is an example. Let A consist of the $n \times n$ matrices with complex entries; if $n \geq 2$, this algebra is not commutative. For any $a \in A$, define

$$\|a\| := \sup\{\|ax\| : x \in \mathbb{C}^n,\ \|x\| \leq 1\},$$

where $\mathbb{C}^n$ is equipped with the Euclidean norm of $\mathbb{R}^{2n}$. This norm then turns A into a finite-dimensional, unital C^*-algebra, and since it is noncommutative, it cannot be of the form $C(K, \mathbb{C})$ for some compact Hausdorff space K.

Every C^*-algebra—commutative or not—can be represented as bounded linear operators on some Hilbert space. Going into the details here would go too far in this book, but loosely speaking it means that every C^*-algebra is an algebra of matrices, which may be "infinitely large" (whatever that may mean).

One can already get a good impression of noncommutative C^*-algebras in the finite-dimensional situation, and this can be done with a surprisingly elementary mathematical toolkit. The text [FARENICK 01], written for undergraduates, is highly recommended. A more advanced introduction to C^*-algebras, requiring a background in complex and functional analysis, is the equally recommended book [MURPHY 90].

5

Basic Algebraic Topology

A grand theme in any mathematical discipline is the classification of its objects: When are two such objects "essentially the same"?

In linear algebra, for example, the objects of study are the finite-dimensional vector spaces. One can agree that two such vector spaces are "essentially the same" if they are isomorphic as linear spaces, and one learns in any introduction to the subject that two finite-dimensional vector spaces (over the same field) are isomorphic if and only if their dimensions coincide. For example, $\mathbb{R}^3$ and the vector space of all real polynomials of degree at most two are isomorphic because they are both three-dimensional, but $\mathbb{R}^3$ and $\mathbb{R}^2$ are not isomorphic because their respective dimensions are different.

The dimension of a finite-dimensional vector space is what's called a numerical *invariant*: a number assigned to each such vector space, which can be used to tell different spaces apart.

It would be nice if a classification of topological spaces could be accomplished with equal simplicity, but this is too much to expect. There is a notion of dimension for topological spaces (we have only encountered zero-dimensional spaces in this book; see Exercise 3.4.10), and there are other numerical invariants for (at least certain) topological spaces. In general, however, mere numbers are far too unstructured to classify objects as diverse as topological spaces.

In algebraic topology, one therefore often does not use numbers, but algebraic objects, mostly groups, as invariants. To each topological space, particular groups are assigned in such a way that, if the spaces are "essentially the same", then so are the associated groups.

5.1 Homotopy and the Fundamental Group

If two topological spaces are homeomorphic, then they are "the same" in the sense that they are indistinguishable as far as every property is concerned that can be formulated in terms of their topologies. Hence, for example, the closed unit disc in $\mathbb{R}^n$, which is compact, cannot be homeomorphic to the open unit disc, which isn't. Very often, however, it is not so straightforward to decide whether two spaces are homeomorphic.

The closed unit disc in $\mathbb{R}^2$ and its boundary $\mathbb{S}^1$ are both compact, connected, and metrizable spaces. Why shouldn't they be homeomorphic? One can show by elementary means that they aren't (see Exercise 1 below), but the argument requires a little trick. And what about the closed unit disc in $\mathbb{R}^2$ and a closed annulus? Again, both spaces are compact, connected, and metrizable, but—unless the annulus is a circle—the trick from Exercise 1 is useless.

We show in this section that the closed unit disc in $\mathbb{R}^2$ cannot be homeomorphic to an annulus (Example 5.1.27 below), but for this purpose we require new and more powerful tools than we have developed so far. As can be expected, developing those tools requires new definitions.

Definition 5.1.1. *Let $(X, \mathcal{T}_X)$ and $(Y, \mathcal{T}_Y)$ be topological spaces. Two continuous maps $f, g \colon X \to Y$ are called* homotopic, $f \sim g$ *in symbols, if there is a continuous map $F \colon [0, 1] \times X \to Y$ such that*

$$F(0, x) = f(x) \quad and \quad F(1, x) = g(x) \qquad (x \in X).$$

The map F is called a homotopy *between f and g.*

Intuitively, one can think of a homotopy as a way of "morphing" one function into another.

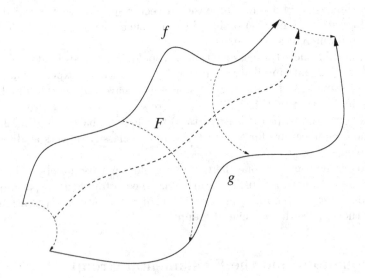

Fig. 5.1: Homotopy

Examples 5.1.2. (a) Let $(X, \mathcal{T})$ be any topological space, and let $C \neq \varnothing$ be a convex subset of a normed space. Then any two continuous maps $f, g : X \to C$ are homotopic. Just let

$$F(t, x) := (1 - t)f(x) + t\,g(x) \qquad (t \in [0, 1], \; x \in X).$$

(b) Let $X = [0, 2\pi]$, and let $Y = \mathbb{S}^2$. Let

$$f : [0, 2\pi] \to \mathbb{S}^2, \quad \theta \mapsto (\cos\theta, \sin\theta, 0)$$

and

$$g : [0, 2\pi] \to \mathbb{S}^2, \qquad \theta \mapsto (0, 0, 1).$$

Define a homotopy between f and g by letting

$$F(t, \theta) = \left(\cos\left(\frac{\pi}{2}t\right)\cos\theta, \cos\left(\frac{\pi}{2}t\right)\sin\theta, \sin\left(\frac{\pi}{2}t\right)\right)$$

for $t \in [0, 1]$ and $\theta \in [0, 2\pi]$.

(c) Let $(X, \mathcal{T}_X)$ be any topological space, let $(Y, \mathcal{T}_Y)$ be totally disconnected, and let $f, g : X \to Y$ be homotopic. Let $F : [0, 1] \times X \to Y$ be a homotopy between f and g. Fix $x \in X$, and define $F_x : [0, 1] \to Y$ by letting $F_x(t) := F(t, x)$. Then F_x is continuous, so that $F_x([0, 1])$ is a connected subspace of Y. Since Y is totally disconnected, this means that $F_x([0, 1])$ consists of one point only, so that

$$f(x) = F(0, x) = F_x(0) = F_x(1) = F(1, x) = g(x).$$

Hence, f and g must be identical.

Definition 5.1.3. *Let $(X, \mathcal{T}_X)$ and $(Y, \mathcal{T}_Y)$ be topological spaces. Then X and Y are called* homotopically equivalent *or of the same homotopy type if there are continuous maps $f : X \to Y$ and $g : Y \to X$ such that $f \circ g \sim \mathrm{id}_Y$ and $g \circ f \sim \mathrm{id}_X$. The maps f and g are then called* homotopy equivalences.

Examples 5.1.4. (a) Any two homeomorphic spaces are trivially homotopically equivalent.

(b) Let C_1 and C_2 be nonempty convex subspaces of normed spaces E_1 and E_2, respectively. Let $f : C_1 \to C_2$ and $g : C_2 \to C_1$ be any continuous maps. In view of Example 5.1.2(a), it is clear that $f \circ g \sim \mathrm{id}_{C_2}$ and $g \circ f \sim \mathrm{id}_{C_1}$, so that C_1 and C_2 are homotopically equivalent. In particular, every convex subset of a normed space is homotopically equivalent to a singleton space. This shows at once that homotopic equivalence is a much weaker notion than homeomorphism.

(c) Let E be a normed space, let $x_0 \in E$, and let $0 \leq r \leq R \leq \infty$. Then the closed *annulus* with center x_0, inner radius r, and outer radius R is the set

$$A_{r,R}[x_0] := \{x \in E : r \leq \|x - x_0\| \leq R\}.$$

(Of course, one can also consider open and half-open annuli.) In the case $r = R < \infty$, the annulus $A_{r,R}[x_0]$ is nothing but the sphere $S_r[x_0]$. Suppose that $0 < r \le \rho \le R$. We claim that $S_\rho[x_0]$ and $A_{r,R}[x_0]$ are homotopically equivalent. Let

$$f \colon A_{r,R}[x_0] \to S_\rho[x_0], \quad x \mapsto \frac{\rho}{\|x - x_0\|}(x - x_0) + x_0,$$

and let $g \colon S_\rho[x_0] \to A_{r,R}[x_0]$ be the inclusion map. It is then clear that $f \circ g = \mathrm{id}_{S_\rho[x_0]}$. Define

$$F(t,x) := \frac{(1-t)\rho + t\|x - x_0\|}{\|x - x_0\|}(x - x_0) + x_0 \qquad (t \in [0,1], \; x \in A_{r,R}[x_0]).$$

Then F is easily seen to be a homotopy between $g \circ f$ and $\mathrm{id}_{A_{r,R}[x_0]}$. Since any two spheres in E are homeomorphic, it follows that $A_{r,R}[x_0]$ and *any* sphere in E are homotopically equivalent.

(d) We claim that the Cantor set C (Example 3.4.18(e)) and $\mathbb{Q}$ are not homotopically equivalent. Assume that there are homotopy equivalences $f \colon C \to \mathbb{Q}$ and $g \colon \mathbb{Q} \to C$. By Example 5.1.2(c) (remember that both C and $\mathbb{Q}$ are totally disconnected) this means that $f \circ g = \mathrm{id}_\mathbb{Q}$ and $g \circ f = \mathrm{id}_C$; that is, C and $\mathbb{Q}$ are even homeomorphic. However, C is compact whereas $\mathbb{Q}$ isn't.

This list is not very impressive. In particular, we still lack satisfactory tools to check if two spaces *fail* to be homotopically equivalent.

Definition 5.1.5. *Let $(X, \mathcal{T})$ be a topological space. Two paths $\gamma, \gamma' \colon [0,1] \to X$ with $\gamma(0) = x_0 = \gamma'(0)$ and $\gamma(1) = x_1 = \gamma'(1)$ are called* path homotopic *($\gamma \simeq \gamma'$ in symbols) if there is a continuous map $\Gamma \colon [0,1] \times [0,1] \to X$, a* path homotopy *between γ and γ', such that*

$$\Gamma(0,s) = \gamma(s) \quad and \quad \Gamma(1,s) = \gamma'(s) \qquad (s \in [0,1])$$

as well as

$$\Gamma(t,0) = x_0 \quad and \quad \Gamma(t,1) = x_1 \qquad (t \in [0,1]).$$

Path homotopy is hence slightly stronger than mere homotopy: while morphing γ into γ', we retain control over $\gamma(0) = \gamma'(0)$ and $\gamma(1) = \gamma'(1)$.

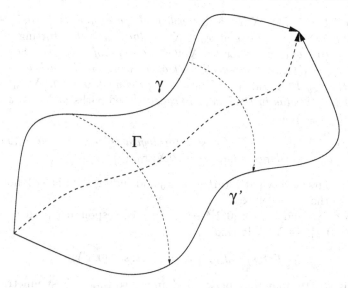

Fig. 5.2: Path homotopy

Examples 5.1.6. (a) Let $C \neq \varnothing$ be a convex subset of a normed space, and let $x_0 \in C$ be arbitrary. For any path γ in C with $\gamma(0) = \gamma(1) = x_0$, we define

$$\Gamma(t, s) := t x_0 + (1 - t)\gamma(s) \qquad (t, s \in [0, 1]).$$

It follows that Γ is a path homotopy between γ and γ' defined as $\gamma'(s) := x_0$ for $s \in [0, 1]$.

(b) Let $X = \mathbb{R}^2 \setminus \{(0, 0)\}$, let

$$\gamma \colon [0, 1] \to X, \quad s \mapsto (\cos(2\pi s), \sin(2\pi s)),$$

and let $\gamma'(s) := (1, 0)$ for $s \in [0, 1]$. Intuitively, it is clear that γ and γ' are *not* path homotopic: γ describes the unit circle, and when we try to "morph" it into the singleton $(1, 0)$ without breaking it up, the origin $(0, 0)$ simply "gets in our way." This, of course, is not a mathematically acceptable proof. We return to this example later on and then give a rigorous proof for the path nonhomotopy of γ and γ'; we don't have the tools yet. It is easy to see, however, that γ and γ' are homotopic in the sense of Definition 5.1.1. Define $F \colon [0, 1]^2 \to X$ through

$$F(t, s) := (\cos(2\pi s(1 - t)), \sin(2\pi s(1 - t))) \qquad (t, s \in [0, 1]).$$

Then F is a homotopy, but not a path homotopy, between γ and γ'.

For convenience, we introduce some more terminology and notation.

Definition 5.1.7. *Let $(X, \mathcal{T})$ be a topological space, and let $\gamma \colon [0,1] \to X$ be a path. We say that γ starts at $x_0 \in X$ (or that x_0 is the starting point) of γ if $x_0 = \gamma(0)$, and we call $x_1 \in X$ the endpoint of γ (or say that γ ends at x_1) if $x_1 = \gamma(1)$. We denote the set of all paths in X starting at x_0 and ending at x_1 by $P(X; x_0, x_1)$. If $x_0 = x_1$, we simply write $P(X, x_0)$ instead of $P(X; x_0, x_1)$; paths in $P(X, x_0)$ are called closed paths or loops, and x_0 is called their base point.*

Proposition 5.1.8. *Let $(X, \mathcal{T})$ be a topological space, and let $x_0, x_1 \in X$. Then $\simeq$ is an equivalence relation on $P(X; x_0, x_1)$.*

Proof. Of course, every path starting at x_0 and ending at x_1 is path homotopic to itself, so that $\simeq$ is reflexive.

Let $\gamma \simeq \gamma'$, and let $\Gamma \colon [0,1]^2 \to X$ be a corresponding path homotopy. Define $\tilde{\Gamma} \colon [0,1]^2 \to X$ by letting

$$\tilde{\Gamma}(t,s) = \Gamma(1-t,s) \qquad (t, s \in [0,1]).$$

Then $\tilde{\Gamma}$ is a path homotopy between γ' and γ, so that $\simeq$ is symmetric.

Let $\gamma \simeq \gamma'$ and let $\gamma' \simeq \gamma''$. Let Γ and Γ' be corresponding path homotopies. Define $\tilde{\Gamma}$ by letting

$$\tilde{\Gamma}(t,s) := \begin{cases} \Gamma(2t,s), & t \in \left[0, \frac{1}{2}\right], \ s \in [0,1], \\ \Gamma'(2t-1,s), & t \in \left[\frac{1}{2}, 1\right], \ s \in [0,1]. \end{cases}$$

Then $\tilde{\Gamma}$ is a path homotopy between γ and γ''. Hence, $\simeq$ is also transitive. $\square$

We can now define the fundamental "group" of a topological space.

Definition 5.1.9. *Let $(X, \mathcal{T})$ be a topological space, and let $x_0 \in X$. Then the set of all equivalence classes of loops in $P(X; x_0)$ with respect to $\simeq$ is called the fundamental group of X at x_0 and is denoted by $\pi_1(X, x_0)$.*

Just calling something a group doesn't make it one yet. In order to show that the fundamental group of a space can indeed be turned into a group, we prove a series of lemmas.

Lemma 5.1.10. *Let $(X, \mathcal{T})$ be a topological space, let $x_0, x_1, x_2 \in X$, and let $\gamma_1, \gamma_1' \in P(X; x_0, x_1)$ and $\gamma_2, \gamma_2' \in P(X; x_1, x_2)$ be such that $\gamma_1 \simeq \gamma_1'$ and $\gamma_2 \simeq \gamma_2'$. Then $\gamma_1 \odot \gamma_2, \gamma_1' \odot \gamma_2' \in P(X; x_0, x_2)$ are also path homotopic.*

Proof. Let $\Gamma_1, \Gamma_2 \colon [0,1] \times [0,1] \to X$ be the path homotopies involved. Define $\tilde{\Gamma} \colon [0,1] \to [0,1] \to X$ as follows,

$$\tilde{\Gamma}(t,s) := \begin{cases} \Gamma_1(t, 2s), & t \in [0,1], \ s \in \left[0, \frac{1}{2}\right], \\ \Gamma_2(t, 2s-1), & t \in [0,1], \ s \in \left[\frac{1}{2}, 1\right]. \end{cases}$$

It is immediate that $\tilde{\Gamma}$ is a path homotopy between $\gamma_1 \odot \gamma_2$ and $\gamma_1' \odot \gamma_2'$. $\square$

Lemma 5.1.11. *Let $(X, \mathcal{T})$ be a topological space, let $x_0, x_1, x_2, x_3 \in X$, and let $\gamma_j \in P(X; x_{j-1}, x_j)$ for $j = 1, 2, 3$. Then $(\gamma_1 \odot \gamma_2) \odot \gamma_3, \gamma_1 \odot (\gamma_2 \odot \gamma_2) \in P(X; x_0, x_3)$ are path homotopic.*

Proof. Define $\Gamma : [0, 1]^2 \to X$ by letting

$$
\Gamma(t, s) := \begin{cases}
\gamma_1\left(\frac{4s}{1+t}\right), & 0 \leq s \leq \frac{t+1}{4}, \\
\gamma_2(4s - 1 - t), & \frac{t+1}{4} \leq s \leq \frac{t+2}{4}, \\
\gamma_3\left(1 - \frac{4(1-s)}{2-t}\right), & \frac{t+2}{4} \leq s \leq 1.
\end{cases}
$$

This definition may look complicated at first glance, but the idea behind it is, in fact, quite simple, as the following sketch shows.

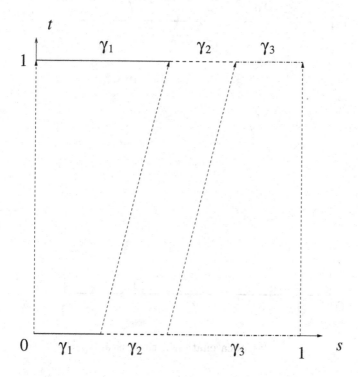

Fig. 5.3: Associativity of path concatenation modulo path homotopy

It is routinely checked that Γ is a path homotopy between $(\gamma_1 \odot \gamma_2) \odot \gamma_3$ and $\gamma_1 \odot (\gamma_2 \odot \gamma_3)$. $\square$

If x is any point in a topological space, we use the same symbol to denote the constant loop with base point x, given by $\gamma(t) := x$ for $t \in [0, 1]$.

Lemma 5.1.12. *Let $(X, \mathcal{T})$ be a topological space, let $x_0, x_1 \in X$, and let $\gamma \in P(X; x_0, x_1)$. Then γ, $x_0 \odot \gamma$, $\gamma \odot x_1 \in P(X; x_0, x_1)$ are path homotopic.*

Proof. Define $\Gamma \colon [0,1]^2 \to X$ through

$$\Gamma(t, s) := \begin{cases} x_0, & 0 \le s \le \tfrac{1}{2}t, \\ \gamma\left(\tfrac{2s-t}{2-t}\right), & \tfrac{1}{2}t \le s \le 1. \end{cases}$$

Then Γ is a path homotopy between γ and $x_0 \odot \gamma$. As in the proof of Lemma 5.1.11, the idea behind the definition of Γ becomes clearer through a sketch.

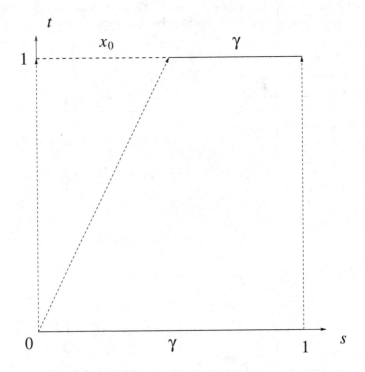

Fig. 5.4: Concatenation with constant paths

In an analogous way, one constructs a path homotopy between γ and $\gamma \odot x_1$. $\square$

Lemma 5.1.13. *Let $(X, \mathcal{T})$ be a topological space, let $x_0, x_1 \in X$, and let $\gamma \in P(X; x_0, x_1)$. Then $\gamma \odot \gamma^{-1} \in P(X, x_0)$ and x_0 as well as $\gamma^{-1} \odot \gamma \in P(X, x_1)$ and x_1 are path homotopic.*

Proof. Define $\Gamma \colon [0,1]^2 \to X$ as follows,

$$\Gamma(t,s) := \begin{cases} \gamma(2s), & 0 \le s \le \frac{1}{2}t, \\ \gamma(t), & \frac{1}{2}t \le s \le 1 - \frac{1}{2}t \\ \gamma(2-2s), & 1 - \frac{1}{2}t \le s \le 1. \end{cases}$$

Again, a sketch illustrates the idea.

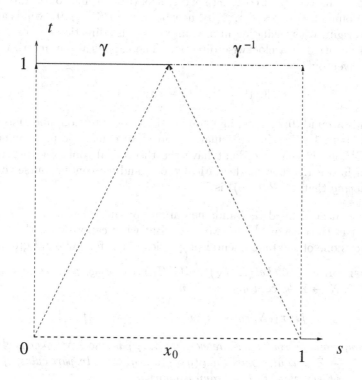

Fig. 5.5: Concatenation with the reversed path

Then Γ is a path homotopy between x_0 and $\gamma \odot \gamma^{-1}$.

Similarly, one obtains a path homotopy between $\gamma^{-1} \odot \gamma$ and x_1. $\square$

Combining Lemmas 5.1.10 to 5.1.13, we obtain the following.

Theorem 5.1.14. *Let $(X, \mathcal{T})$ be a topological space, and let $x_0 \in X$. Then $\pi_1(X, x_0)$ is a group under the operation*

$$[\gamma_1] \cdot [\gamma_2] := [\gamma_1 \odot \gamma_2] \qquad ([\gamma_1], [\gamma_2] \in \pi_1(X, x_0)),$$

where $[\gamma] \in \pi_1(X, x_0)$ denotes the equivalence class of $\gamma \in P(X, x_0)$.

Proof. In view of Lemma 5.1.10, $\cdot$ is well defined.

Associativity of $\cdot$ follows from Lemma 5.1.11, $[x_0]$ is the neutral element of $(\pi_1(X, x_0), \cdot)$ by Lemma 5.1.12, and by Lemma 5.1.13, each $[\gamma] \in \pi_1(X, x_0)$ has an inverse, namely $[\gamma^{-1}]$.

All in all, $\pi_1(X, x_0)$ is indeed a group. $\square$

Examples 5.1.15. (a) Let C be a convex subset of a normed space, and let $x_0 \in C$. By Example 5.1.6(a), it is clear that $\pi_1(C, x_0) = \{[x_0]\} \cong \{0\}$.

(b) We now turn to the fundamental group of the unit circle $\mathbb{S}^1$. We compute $\pi_1(\mathbb{S}^1, (1, 0))$ at the end of the next section (Example 5.2.7 below) with the help of the theory of covering spaces: it is (isomorphic to) $\mathbb{Z}$. Intuitively, one counts for $[\gamma] \in \pi_1(\mathbb{S}^1, (1, 0))$ how often $\gamma \in P(\mathbb{S}^1, (1, 0))$ winds around the origin, where winding in a counterclockwise direction counts positive and negative in a clockwise direction. The canonical parametrization of $\mathbb{S}^1$, given by

$$\gamma\colon [0, 1] \to \mathbb{S}^1, \quad t \mapsto (\cos(2\pi t), \sin(2\pi t))$$

winds around $(0, 0)$ once, in a counterclockwise direction, and thus corresponds to $1 \in \mathbb{Z}$. (In particular, γ and $(1, 0)$ cannot be path homotopic in $\mathbb{S}^1$.) Right now, we don't have the theoretical tools yet to turn this idea into a rigorous mathematical proof, and we content ourselves with *believing* that $\pi_1(\mathbb{S}^1, (1, 0))$ is $\mathbb{Z}$.

So, we have defined the fundamental group, and calculated it—well, not quite yet in the case of $\mathbb{S}^1$—in two cases. But what can we do with it?

Here is one of the fundamental properties of the fundamental group.

Proposition 5.1.16. *Let $(X, \mathcal{T}_X)$ and $(Y, \mathcal{T}_Y)$ be topological spaces, let $x_0 \in X$, and let $f\colon X \to Y$ be continuous. Then*

$$f_*\colon \pi_1(X, x_0) \to \pi_1(Y, f(x_0)), \quad [\gamma] \mapsto [f \circ \gamma]$$

is a group homomorphism. Moreover, if $(Z, \mathcal{T}_Z)$ is another topological space, and $g\colon Y \to Z$ is continuous, then $(g \circ f)_* = g_* \circ f_*$. In particular, if f is a homeomorphism, then f_* is a group isomorphism.

Proof. Straightforward. □

Example 5.1.17. Let C be any nonempty convex subset of a normed space. Then C and $\mathbb{S}^1$, the unit sphere in $\mathbb{R}^2$, cannot be homeomorphic. Otherwise, $\pi_1(C, x_0)$ and $\pi_1(\mathbb{S}^1, (1, 0))$ would be isomorphic for some $x_0 \in C$, and this is impossible. In particular, $\mathbb{S}^1$ and the closed unit ball in $\mathbb{R}^2$ are not homeomorphic (for an alternative, more elementary proof of this see Exercise 1 below).

For another application of the fundamental group, we present another definition.

Definition 5.1.18. *Let $(X, \mathcal{T})$ be a topological space. Then a subspace Y of X is called a retract if there is a retraction of Y: a continuous map $r\colon X \to Y$ which is the identity on Y.*

Example 5.1.19. The unit sphere $\mathbb{S}^{n-1}$ is a retract of $\mathbb{R}^n \setminus \{0\}$. A retraction is given by

$$r \colon \mathbb{R}^n \setminus \{0\} \to \mathbb{S}^{n-1}, \quad x \mapsto \frac{x}{\|x\|}.$$

Proposition 5.1.20. *Let $(X, \mathcal{T})$ be a topological space, let Y be a subspace of X, let $r \colon X \to Y$ be a retraction of Y, and let $y_0 \in Y$. Then the group homomorphism $\iota_* \colon \pi_1(Y, y_0) \to \pi_1(X, y_0)$, where $\iota \colon Y \to X$ is the canonical inclusion, is injective.*

Proof. Let $r \colon X \to Y$ be a retraction of Y; that is, r is a continuous left inverse of ι. Consequently, r_* is a left inverse of ι_*. Hence, ι_* is injective. □

Examples 5.1.21. (a) In Example 5.1.6(b), we claimed that the paths

$$\gamma \colon [0,1] \to \mathbb{R}^2 \setminus \{(0,0)\}, \quad t \mapsto (\cos(2\pi t), \sin(2\pi t))$$

and $(1,0)$ were not path homotopic in $\mathbb{R}^2 \setminus \{(0,0)\}$ because $(0,0)$ somehow "was in the way." In terms of the fundamental group $\pi_1(\mathbb{R}^2 \setminus \{(0,0)\}, (1,0))$, another way of wording this is that $[\gamma]$ and $[(1,0)]$ are different elements of $\pi_1(\mathbb{R}^2 \setminus \{(0,0)\}, (1,0))$. With what we have learned so far about fundamental groups, we can now give a rigorous argument for our claim (with one gap that is closed later). Since $\gamma([0,1]) \subset \mathbb{S}^1$, the path γ also yields an element of $\pi_1(\mathbb{S}^1, (1,0))$, which we denote by $[\gamma]_{\mathbb{S}^1}$ to tell it apart from $[\gamma] \in \pi_1(\mathbb{R}^2 \setminus \{(0,0)\}, (1,0))$; similarly, we write $[(1,0)]_{\mathbb{S}^1}$ for the equivalence class of $(1,0)$ in $P(\mathbb{S}^1, (1,0))$. In Example 5.1.15(b), we had convinced ourselves (admittedly with a lot of handwaving) that $[\gamma]_{\mathbb{S}^1} \neq [(1,0)]_{\mathbb{S}^1}$. Since $\mathbb{S}^1$ is a retract of $\mathbb{R}^2 \setminus \{(0,0)\}$, the canonical map from $\pi_1(\mathbb{S}^1, (1,0))$ to $\pi_1(\mathbb{R}^2 \setminus \{(0,0)\}, (1,0))$ is injective, so that $[\gamma] \neq [(1,0)]$, as claimed in Example 5.1.6(b). (This "proof," of course, depends on some yet unproven claims about $\pi_1(\mathbb{S}^1, (1,0))$: they are proven fully rigorously in Example 5.2.7 below.)

(b) If we believe that $\pi_1(\mathbb{S}^1, (1,0))$ is nonzero (Example 5.1.15(b)), then $\mathbb{S}^1$ cannot be homeomorphic to any retract of a nonempty convex subset of some normed linear space. In particular, $\mathbb{S}^1$ is not a retract of the closed unit ball of $\mathbb{R}^2$.

This last example has a lovely application.

Theorem 5.1.22 (Brouwer's fixed point theorem for $n = 1, 2$). *Let $n = 1, 2$, and let B_n denote the closed unit ball in $\mathbb{R}^n$. Then every continuous map $f \colon B_n \to B_n$ has a fixed point.*

Proof. Assume that f does not have any fixed point; that is, $f(x) \neq x$ for all $x \in B_n$. For each $x \in B_n$, let $r(x)$ denote the unique intersection point of the line starting at $f(x)$ and passing through x (in this direction!) with $\mathbb{S}^{n-1}$.

It is immediate that r is continuous such that, if $x \in \mathbb{S}^{n-1}$, we have $r(x) = x$. Hence, r is a retraction of $\mathbb{S}^{n-1}$. This part of the proof, by the way, does not require at all that $n = 1, 2$.

If $n = 1$, this is impossible because then $\mathbb{S}^0 = \{-1, 1\} = r([-1, 1])$ would have to be connected. If $n = 2$, this is impossible by Example 5.1.21(b).   □

The following sketch illustrates the idea of the proof.

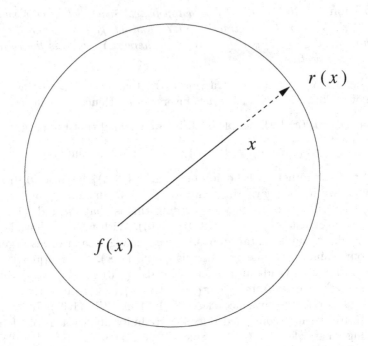

Fig. 5.6: Proof of Brouwer's fixed point theorem

With more sophisticated methods—which we do not develop in this book—it can be shown that $\mathbb{S}^{n-1}$ is not a retract of B_n for *every* $n \in \mathbb{N}$, so that Brouwer's fixed point theorem holds in all dimensions.

Toward the end of this section, we focus on two further questions around the fundamental group. The first one is: how does it depend on the base point of the loops. Given a space $(X, \mathcal{T})$ and $x_0, x_1 \in X$, how are $\pi_1(X, x_0)$ and $\pi_1(X, x_1)$ related? In general, nothing can be said unless x_0 and x_1 lie in the same component of X (Exercise 6 below). If x_0 and x_1, however, not only lie in the same component, but can be connected by a path, the situation is different.

Proposition 5.1.23. *Let $(X, \mathcal{T})$ be a topological space, and let $x_0, x_1 \in X$ be such that there is $\alpha \in P(X; x_0, x_1)$. Then*

$$\phi_\alpha \colon \pi_1(X, x_0) \to \pi_1(X, x_1), \quad [\gamma] \mapsto [\alpha^{-1} \odot \gamma \odot \alpha]$$

is a group isomorphism.

Proof. It follows from Lemma 5.1.10 that ϕ_α is well defined.

Invoking Lemmas 5.1.11, 5.1.13, and then 5.1.12, we obtain that ϕ_α is indeed a group homomorphism, whose inverse is given by

$$\pi_1(X, x_1) \to \pi_1(X, x_0), \quad [\gamma] \mapsto [\alpha \odot \gamma \odot \alpha^{-1}].$$

This completes the proof. $\square$

Corollary 5.1.24. *Let* $(X, \mathcal{T}_X)$ *be a path connected topological space. Then* $\pi_1(X, x_0)$ *and* $\pi_1(X, x_1)$ *are isomorphic for any* $x_0, x_1 \in X$.

Even though $\pi_1(X, x_0)$ and $\pi_1(X, x_1)$ are isomorphic for path connected X, no matter how x_0 and x_1 are chosen, the isomorphism need not be canonical: it depends on the choice for a path connecting x_0 and x_1.

In Proposition 5.1.16, we saw that any continuous map f between two topological spaces induces a homomorphism f_* of fundamental groups. Our last theorem in this section tells us how f_* and g_* are related when f and g are both continuous and homotopic.

Theorem 5.1.25. *Let* $(X, \mathcal{T}_X)$ *and* $(Y, \mathcal{T}_Y)$ *be topological spaces, let* $x_0 \in X$, *and let* $f, g \colon X \to Y$ *be homotopic. Then:*

(i) *If* $F \colon [0, 1] \times X \to Y$ *is a homotopy between* f *and* g, *then*

$$\alpha \colon [0, 1] \to Y, \quad t \mapsto F(t, x_0)$$

is a path in $P(Y; f(x_0), g(x_0))$.

(ii) *The diagram*

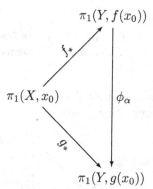

commutes, where ϕ_α *is defined as in Proposition 5.1.23.*

Proof. (i) is obvious.

For (ii), we need to show that

$$\alpha \odot (g \circ \gamma) \simeq (f \circ \gamma) \odot \alpha \qquad (\gamma \in P(X, x_0)).$$

Let $\gamma \in P(X, x_0)$, and define $\gamma_0, \gamma_1 \colon [0, 1] \to [0, 1] \times X$ through

$$\gamma_0(t) := (0, \gamma(t)) \quad \text{and} \quad \gamma_1(t) := (1, \gamma(t)) \qquad (t \in [0, 1]).$$

Furthermore, let

$$c \colon [0, 1] \to [0, 1] \times X, \quad t \mapsto (t, x_0).$$

It is then immediate that

$$F \circ \gamma_0 = f \circ \gamma, \quad F \circ \gamma_1 = g \circ \gamma, \quad \text{and} \quad F \circ c = \alpha.$$

Define

$$G \colon [0, 1] \times [0, 1] \to [0, 1] \times X, \quad (t, s) \mapsto (t, \gamma(s)),$$

and let $\beta_1, \ldots, \beta_4$ be parametrizations of the line segments that make up $\partial([0, 1] \times [0, 1])$; that is,

$$\beta_1(t) := (0, t), \quad \beta_2(t) := (t, 1), \quad \beta_3(t) := (1, t), \quad \text{and} \quad \beta_4(t) := (t, 0)$$

for $t \in [0, 1]$. It is immediate that

$$G \circ \beta_1 = \gamma_0, \quad G \circ \beta_3 = \gamma_1, \quad \text{and} \quad G \circ \beta_2 = G \circ \beta_4 = c.$$

The concatenated paths $\beta_1 \odot \beta_2$ and $\beta_4 \odot \beta_3$ are in $[0, 1]^2$, which is a convex subset of $\mathbb{R}^2$, so that they are path homotopic, via a path homotopy, say B. It follows that $G \circ B$ is a path homotopy in $[0, 1] \times X$ between $\gamma_0 \odot c$ and $c \odot \gamma_1$, and, consequently, that $F \circ (G \circ B)$ is a path homotopy between

$$(F \circ \gamma_0) \odot (F \circ c) = (f \circ \gamma) \odot \alpha \quad \text{and} \quad (F \circ c) \odot (F \circ \gamma_1) = \alpha \odot (g \circ \gamma).$$

This proves (ii). □

Corollary 5.1.26. *Let $(X, \mathcal{T}_X)$ and $(Y, \mathcal{T}_Y)$ be topological spaces, let $x_0 \in X$, and let $f \colon X \to Y$ be a homotopy equivalence. Then $f_* \colon \pi_1(X, x_0) \to \pi_1(Y, f(x_0))$ is a group isomorphism.*

Proof. Let $g \colon Y \to X$ be such that $f \circ g \sim \mathrm{id}_Y$ and $g \circ f \sim \mathrm{id}_X$. By Theorem 5.1.25, we obtain a commutative diagram

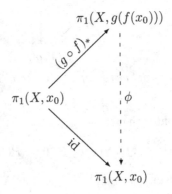

for a suitable group isomorphism ϕ; that is, id $= \phi \circ (g \circ f)_* = (\phi \circ g_*) \circ f_*$. In particular, $f_* : \pi_1(X, x_0) \to \pi_1(Y, f(x_0))$ has a left inverse and thus is injective.

In a similar vein, we obtain a group isomorphism $\psi : \pi_1(Y, f(g(f(x_0)))) \to \pi_1(Y, f(x_0))$ such that

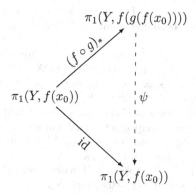

commutes; that is, id $= \psi \circ (f \circ g)_* = (\psi \circ f_*) \circ g_*$. Hence, $\psi \circ f_* : \pi_1(X, g(f(x_0))) \to \pi_1(Y, f(x_0))$ has a right inverse and thus is surjective, and since ψ is bijective, $f_* : \pi_1(X, g(f(x_0))) \to \pi_1(Y, f(g(f(x_0))))$ must be surjective, too. Since $g \circ f \sim \mathrm{id}_X$, there is $\alpha \in P(X; g(f(x_0)), x_0)$— see Theorem 5.1.25(i)—so that $f \circ \alpha \in P(X; f(g(f(x_0))), f(x_0))$. With $\phi_\alpha : \pi_1(X, g(f(x_0))) \to \pi_1(X, x_0)$ and $\phi_{f\circ\alpha} : \pi_1(Y, f(g(f(x_0)))) \to \pi_1(Y, f(x_0))$ as in Proposition 5.1.23, we obtain a commutative diagram

$$
\begin{array}{ccc}
\pi_1(X, g(f(x_0))) & \xrightarrow{f_*} & \pi_1(Y, f(g(f(x_0)))) \\
{\scriptstyle \phi_\alpha}\downarrow & & \downarrow{\scriptstyle \phi_{f\circ\alpha}} \\
\pi_1(X, x_0) & \xrightarrow{\;\;f_*\;\;} & \pi_1(Y, f(x_0)).
\end{array}
$$

Since ϕ_α and $\phi_{f\circ\alpha}$ are group *isomorphisms*, it follows that $f_* : \pi_1(X, x_0) \to \pi_1(Y, f(x_0))$ is also surjective.

All in all, $f_* : \pi_1(X, x_0) \to \pi_1(Y, f(x_0))$ is bijective and therefore a group isomorphism. $\square$

Example 5.1.27. Let A be any closed annulus in $\mathbb{R}^2$ whose inner radius is strictly positive, and let $x_0 \in A$. By Example 5.1.4(c), A and $\mathbb{S}^1$ are of the same homotopy type, so that $\pi_1(A, x_0) \cong \mathbb{Z}$ by Corollaries 5.1.26 and 5.1.24. Invoking Corollary 5.1.26 again, we see that, if C is any nonempty convex subset of a normed space, then A and C cannot be of the same homotopy type. In particular, A and the closed unit ball in $\mathbb{R}^2$ are not homotopically equivalent and thus not homeomorphic.

Exercises

1. Prove by elementary means (i.e., without involving any notion of homotopy) that B_2 and $\mathbb{S}^1$ are not homeomorphic. (*Hint*: What can you say about the connectedness of B_2 and $\mathbb{S}^1$ if two distinct points have been removed from both spaces?)

2. Let $(X, \mathcal{T}_X)$, $(Y, \mathcal{T}_Y)$, and $(Z, \mathcal{T}_Z)$ be topological spaces, and let $g, g' : X \to Y$ and $f, f' : Y \to Z$ be continuous maps such that $g \sim g'$ and $f \sim f'$. Show that $f \circ g \sim f' \circ g'$.

3. Let $(X, \mathcal{T}_X)$ and $(Y, \mathcal{T}_Y)$ be topological spaces. Show that $\sim$ is an equivalence relation on the set of all continuous maps from X to Y.

4. A subset S of a normed space is called *star-shaped* with center $x_0 \in S$ if $tx + (1-t)x_0 \in S$ for all $t \in [0,1]$ and $x \in S$.
   (a) Show that every convex set is star-shaped, but that the converse fails.
   (b) Show that every star-shaped set is path connected.
   (c) Let S be star-shaped with center x_0, and let $\gamma : [0,1] \to S$ be a path with $\gamma(0) = \gamma(1) = x_0$. Show that $\gamma \simeq x_0$.

5. Let $(X, \mathcal{T}_X)$ and $(Y, \mathcal{T}_Y)$ be topological spaces, let $x_0 \in X$ and $y_0 \in Y$, let $X \times Y$ be equipped with the product topology, and let $\pi^X : X \times Y \to X$ and $\pi^Y : X \times Y \to Y$ be the canonical projections. Show that

$$\pi_*^X \times \pi_*^Y : \pi_1(X \times Y, (x_0, y_0)) \to \pi_1(X, x_0) \times \pi_1(Y, y_0), \quad [\gamma] \mapsto \left(\pi_*^X([\gamma]), \pi_*^Y([\gamma])\right)$$

   is a group isomorphism.

6. Let $(X, \mathcal{T})$ be a topological space, let $x_0 \in X$, and let Y_{x_0} denote the component of X containing x_0. Show that the inclusion of Y_{x_0} in X induces a group isomorphism of $\pi_1(Y_{x_0}, x_0)$ and $\pi_1(X, x_0)$.

7. Let G be a topological group with identity element e. For $\gamma_1, \gamma_2 \in P(G, e)$, define

$$\gamma_1 \boxdot \gamma_2 : [0,1] \to G, \quad t \mapsto \gamma_1(t)\gamma_2(t).$$

   (a) For $\gamma_1, \gamma_1', \gamma_2, \gamma_2' \in P(G, e)$ such that $\gamma_1 \simeq \gamma_1'$ and $\gamma_2 \simeq \gamma_2'$, show that $\gamma_1 \boxdot \gamma_2 \simeq \gamma_1' \boxdot \gamma_2'$.
   (b) Show that

$$\gamma_1 \odot \gamma_2 \simeq \gamma_1 \boxdot \gamma_2 \simeq \gamma_2 \odot \gamma_1 \qquad (\gamma_1, \gamma_2 \in P(G, e)).$$

   (*Hint*: What is $(\gamma_1 \odot e) \boxdot (e \odot \gamma_2)$ for $\gamma_1, \gamma_2 \in P(G, e)$?)
   (c) Conclude that $\pi_1(G, e)$ is abelian.

5.2 Covering Spaces

In the previous section, we defined the fundamental group of a space, and we showed that it is trivial for certain spaces. Nevertheless, we haven't computed a single nontrivial fundamental group yet (Example 5.1.15(b) was more a heuristic argument than a rigorous computation).

   In this section, we rigorously prove that $\pi_1(\mathbb{S}^1, (1,0))$ is indeed (isomorphic to) $\mathbb{Z}$ as claimed in Example 5.1.15(b). For this purpose, we develop the theory of covering spaces to a minuscule extent.

Definition 5.2.1. *Let $(X, \mathcal{T})$ be a topological space. Then a covering space of X is a pair $\left(\left(\tilde{X}, \tilde{\mathcal{T}} \right), p \right)$ such that:*

(a) $\left(\tilde{X}, \tilde{\mathcal{T}} \right)$ *is a topological space;*

(b) $p \colon \tilde{X} \to X$ *is surjective and continuous;*

(c) *Each $x \in X$ has an open neighborhood U such that $p^{-1}(U)$ is the disjoint union of a family $\mathcal{V}$ of open subsets of $\tilde{X}$ such that $p|_V$ is a homeomorphism onto U for each $V \in \mathcal{V}$.*

The map p is called a covering map *and the elements of $\mathcal{V}$ are called the* sheets *of $p^{-1}(U)$.*

Intuitively, the sheets can be thought of as covering U (hence the name).

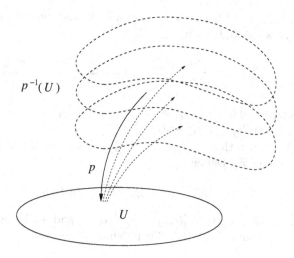

$p^{-1}(U)$

p

U

Fig. 5.7: Covering space

We are mainly interested in the following example.

Example 5.2.2. Consider,

$$p \colon \mathbb{R} \to \mathbb{S}^1, \quad x \mapsto (\cos(2\pi x), \sin(2\pi x)).$$

Let $(x_0, y_0) \in \mathbb{S}^1$. We only consider the case where $x_0 > 0$ (all other cases—$x_0 < 0$, $y_0 > 0$, and $y_0 < 0$—can be treated analogously). Then $U := \{(x, y) \in \mathbb{S}^1 : x > 0\}$ is an open neighborhood of (x_0, y_0) such that

$$p^{-1}(U) = \bigcup \left\{ \left(n - \frac{1}{4}, n + \frac{1}{4} \right) : n \in \mathbb{Z} \right\}.$$

It is easy to see that p maps $\left(n - \frac{1}{4}, n + \frac{1}{4} \right)$ homeomorphically onto U for each $n \in \mathbb{Z}$.

What makes covering spaces interesting for our purpose is that paths can be "lifted" from a given topological space to a covering space:

Lemma 5.2.3. *Let* $(X, \mathcal{T})$ *be a topological space, let* $\left(\left(\tilde{X}, \tilde{\mathcal{T}} \right), p \right)$ *be a covering space of* X, *and let* $x_0 \in X$ *and* $\tilde{x}_0 \in \tilde{X}$ *be such that* $p(\tilde{x}_0) = x_0$. *Then for each path* γ *in* X *with starting point* x_0, *there is a unique path* $\tilde{\gamma}$ *in* $\tilde{X}$ *with starting point* $\tilde{x}_0$ *such that* $p \circ \tilde{\gamma} = \gamma$.

Proof. For each $x \in X$, let U_x be an open neighborhood as specified in Definition 5.2.1. Then $\{\gamma^{-1}(U_x) : x \in X\}$ is an open cover of $[0, 1]$. For each $s \in [0, 1]$, there are $x_s \in X$ and $a_s < b_s$ such that

$$s \in [0, 1] \cap (a_s, b_s) \subset [0, 1] \cap [a_s, b_s] \subset \gamma^{-1}(U_{x_s}).$$

Since $\{(a_s, b_s) : s \in [0, 1]\}$ is an open cover for $[0, 1]$, and since $[0, 1]$ is compact, there are $s_1, \ldots, s_m \in [0, 1]$ with

$$[0, 1] \subset (a_{s_1}, b_{s_1}) \cup \cdots \cup (a_{s_m}, b_{s_m}) \subset \underbrace{[a_{s_1}, b_{s_1}]}_{\subset \gamma^{-1}(U_{x_{s_1}})} \cup \cdots \cup \underbrace{[a_{s_m}, b_{s_m}]}_{\subset \gamma^{-1}(U_{x_{s_m}})} .$$

We may therefore find $0 = t_0 < t_1 < \cdots < t_n = 1$ such that, for $j = 1, \ldots, n$, there is $x_j \in X$ with $[t_{j-1}, t_j] \subset \gamma^{-1}(U_{x_j})$.

Let $x_1 \in X$ be such that $[t_0, t_1] \subset \gamma^{-1}(U_{x_1})$. Let V_1 be the unique sheet of $p^{-1}(U_{x_1})$ containing $\tilde{x}_0$. Define

$$\tilde{\gamma}(t) := (p|_{V_1})^{-1}(\gamma(t)) \qquad (t \in [t_0, t_1]).$$

Now, let $x_2 \in X$ be such that $[t_1, t_2] \subset \gamma^{-1}(U_{x_2})$, and let V_2 be the unique sheet of $p^{-1}(U_{x_2})$ containing $\tilde{\gamma}(t_1)$. Then define

$$\tilde{\gamma}(t) := (p|_{V_2})^{-1}(\gamma(t)) \qquad (t \in [t_1, t_2]).$$

Then, choose $x_3 \in X$ with $[t_2, t_3] \subset \gamma^{-1}(U_{x_3})$ and continue in this fashion.

Successively, we thus obtain a path $\tilde{\gamma}$ in $\tilde{X}$ with $\tilde{\gamma}(0) = \tilde{x}_0$ and $p \circ \tilde{\gamma} = \gamma$.

Suppose that $\tilde{\gamma}'$ is another path with these two properties. Let $\mathcal{V}_1$ be the collection of sheets of $p^{-1}(U_{x_1})$. Since $\gamma([t_0, t_1]) \subset U_{x_1}$, it follows that

$$\tilde{\gamma}'([t_0, t_1]) \subset p^{-1}(U_{x_1}) = \bigcup \{V : V \in \mathcal{V}_1\}.$$

Since $\tilde{\gamma}'([t_0, t_1])$ is connected, and since the sheets of $p^{-1}(U_{x_1})$ are clopen in $p^{-1}(U_{x_1})$, it follows that there is one sheet, say $V_1' \in \mathcal{V}_1$ with $\tilde{\gamma}'([t_0, t_1]) \subset V_1'$. Since $\tilde{\gamma}'(0) = \tilde{x}_0$, it is clear that $V_1' = V_1$, and since $p \circ \tilde{\gamma}' = \gamma$, we see that

$$\tilde{\gamma}'(t) = (p|_{V_1})^{-1}(\gamma(t)) = \tilde{\gamma}(t) \qquad (t \in [t_0, t_1]).$$

Successively, we obtain that $\tilde{\gamma}'(t) = \tilde{\gamma}(t)$ for all $t \in [0, 1]$. $\square$

The following sketch conveys the idea of the proof.

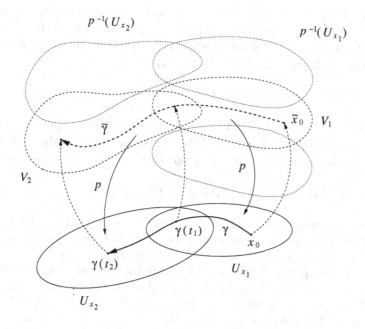

Fig. 5.8: Lifting a path

We call the path $\tilde{\gamma}$ in Lemma 5.2.3, the *lifting* of γ with starting point $\tilde{x}_0$. So, we can lift paths to covering spaces. What about path homotopies?

Lemma 5.2.4. *Let $(X, \mathcal{T})$ be a topological space, let $\left(\left(\tilde{X}, \tilde{\mathcal{T}}\right), p\right)$ be a covering space of X, and let $x_0 \in X$ and $\tilde{x}_0 \in \tilde{X}$ be such that $p(\tilde{x}_0) = x_0$. Then, for each continuous map $\Gamma: [0,1]^2 \to X$ with $\Gamma(0,0) = x_0$, there is a unique continuous map $\tilde{\Gamma}: [0,1]^2 \to \tilde{X}$ with $\tilde{\Gamma}(0,0) = \tilde{x}_0$ and $p \circ \tilde{\Gamma} = \Gamma$. Moreover, if Γ is a path homotopy, then so is $\tilde{\Gamma}$.*

Proof. For each $x \in X$, choose an open neighborhood U_x of x as specified in Definition 5.2.1. An argument similar to that in the proof of Lemma 5.2.3 yields $0 = t_0 < t_1 < \cdots < t_n = 1$ and $0 = s_0 < s_1 < \cdots < s_m = 1$ such that, for any $j \in \{1, \ldots, n\}$ and $k \in \{1, \ldots, m\}$, there is $x_{j,k} \in X$ with $\Gamma([t_{j-1}, t_j] \times [s_{k-1}, s_k]) \subset U_{x_{j,k}}$.

Let $V_{1,1}$ be the sheet of $p^{-1}(U_{1,1})$ containing $\tilde{x}_0$, and define

$$\tilde{\Gamma}(t, s) := (p|_{V_{1,1}})^{-1}(\Gamma(t, s)) \qquad ((t, s) \in [t_0, t_1] \times [s_0, s_1]).$$

Let $V_{2,1}$ be the sheet of $p^{-1}(U_{x_{2,1}})$ containing $\tilde{\Gamma}(t_1, s_0)$. Since sheets are clopen, and since $\tilde{\Gamma}(t_1, [s_0, s_1])$ is connected, we conclude that $\tilde{\Gamma}(t_1, [s_0, s_1]) \subset V_{2,1}$. Defining

$$\tilde{\Gamma}(t,s) := (p|_{V_{2,1}})^{-1}(\Gamma(t,s)) \qquad ((t,s) \in [t_1,t_2] \times [s_0,s_1])$$

thus extends $\tilde{\Gamma}$ to $[t_0,t_2] \times [s_0,s_1]$ as a continuous map. Continuing in this fashion, we obtain a continuous $\tilde{\Gamma}\colon [0,1] \times [s_0,s_1] \to \tilde{X}$ with $\tilde{\Gamma}(0,0) = \tilde{x}_0$.

Next, let $V_{1,2}$ be the sheet of $p^{-1}(U_{x_{1,2}})$ containing $\tilde{\Gamma}(t_0,s_1)$. Again the clopenness of sheets yields that $\tilde{\Gamma}([t_0,t_1],s_1) \subset V_{1,2}$, so that

$$\tilde{\Gamma}(t,s) := (p|_{V_{1,2}})^{-1}(\Gamma(t,s)) \qquad ((t,s) \in [t_0,t_1] \times [s_1,s_2])$$

extends $\tilde{\Gamma}$ continuous to $([0,1] \times [s_0,s_1]) \cup ([t_0,t_1] \times [s_1,s_2])$. Let $V_{2,2}$ be the sheet of $p^{-1}(U_{x_{2,2}})$ containing $\tilde{\Gamma}(t_1,s_1)$. Again a connectedness argument shows that both $\tilde{\Gamma}(t_1,[s_1,s_2]) \subset V_{2,2}$ and $\tilde{\Gamma}([t_1,t_2],s_1) \subset V_{2,2}$, so that

$$\tilde{\Gamma}(t,s) := (p|_{V_{2,2}})^{-1}(\Gamma(t,s)) \qquad ((t,s) \in [t_1,t_2] \times [s_1,s_2])$$

defines a continuous extension of $\tilde{\Gamma}$ to $([0,1] \times [s_0,s_1]) \cup ([t_0,t_2] \times [s_1,s_2])$. Repeating the argument again and again, we eventually obtain a continuous map $\tilde{\Gamma}\colon [0,1] \times [s_0,s_2] \to \tilde{X}$ with $p \circ \tilde{\Gamma} = \Gamma$ and $\tilde{\Gamma}(0,0) = \tilde{x}_0$.

The next step is then to extend $\tilde{\Gamma}$ (using the same arguments as before) to $[0,1] \times [s_0,s_3]$, then to $[0,1] \times [s_0,s_4]$, and so on, till we have it defined on all of $[0,1] \times [0,1]$. This proves the existence of $\tilde{\Gamma}$.

To prove the uniqueness, suppose that $\tilde{\Gamma}'\colon [0,1]^2 \to \tilde{X}$ is any continuous map with $p \circ \tilde{\Gamma}' = \Gamma$ and $\tilde{\Gamma}'(0,0) = \tilde{x}_0$. Let $V_{1,1}$ be the sheet of $p^{-1}(U_{x_{1,1}})$ containing $\tilde{x}_0$. Since $\tilde{\Gamma}'([t_0,t_1] \times [s_0,s_1])$ is connected, it follows that $\tilde{\Gamma}'([t_0,t_1] \times [s_0,s_1]) \subset V_{1,1}$, so that

$$\tilde{\Gamma}'(t,s) = (p|_{V_{1,1}})^{-1}(\Gamma(t,s)) = \tilde{\Gamma}(s,t) \qquad ((t,s) \in [t_0,t_1] \times [s_0,s_1]).$$

Hence, $\tilde{\Gamma}'$ and $\tilde{\Gamma}$ coincide on $[t_0,t_1] \times [s_0,s_1]$. Successively (as in the proof of the existence part) we show that $\tilde{\Gamma}'$ and $\tilde{\Gamma}$ coincide on all of $[0,1] \times [s_0,s_1]$, then on $[0,1] \times [s_0,s_2]$, and so on, and finally on all of $[0,1] \times [0,1]$.

Finally, suppose that Γ is a path homotopy; that is, the maps

$$[0,1] \to X, \quad t \mapsto \Gamma(t,s)$$

are constant for $s \in \{0,1\}$. The uniqueness part of Lemma 5.2.3 then yields that

$$[0,1] \to X, \quad t \mapsto \tilde{\Gamma}(t,s)$$

are also constant for $s \in \{0,1\}$, so that $\tilde{\Gamma}$ is a path homotopy. $\square$

It should be noted that Lemma 5.2.4 contains Lemma 5.2.3 as a particular case, and relies on Lemma 5.2.3 only for the proof of the statement about path homotopies.

A simple consequence of Lemma 5.2.4 is as follows.

Corollary 5.2.5. *Let $(X, \mathcal{T})$ be a topological space, let $\left(\left(\tilde{X}, \tilde{\mathcal{T}}\right), p\right)$ be a covering space of X, let $x_0 \in X$ and $\tilde{x}_0 \in \tilde{X}$ be such that $p(\tilde{x}_0) = x_0$, and let γ and γ' be path homotopic paths in X with starting point x_0. Then their liftings $\tilde{\gamma}$ and $\tilde{\gamma}'$ with starting point $\tilde{x}_0$ are also path homotopic.*

Theorem 5.2.6. *Let $(X, \mathcal{T})$ be a topological space, let $\left(\left(\tilde{X}, \tilde{\mathcal{T}}\right), p\right)$ be a covering space of X, and let $x_0 \in X$ and $\tilde{x}_0 \in \tilde{X}$ be such that $p(\tilde{x}_0) = x_0$. Then:*

(i) *The* lifting correspondence

$$\phi \colon \pi_1(X, x_0) \to p^{-1}(\{x_0\}), \quad [\gamma] \mapsto \tilde{\gamma}(1)$$

is well defined;
(ii) *If $\tilde{X}$ is path connected, then ϕ is surjective;*
(iii) *If $\tilde{X}$ is simply connected, that is, $\tilde{X}$ is path connected and $\pi_1\left(\tilde{X}, \tilde{x}_0\right) = \{0\}$ holds, then ϕ is bijective.*

Proof. Let $\gamma, \gamma' \in P(X, x_0)$ be path homotopic. Then so are their liftings by Corollary 5.2.5 and thus, in particular, have the same endpoints. This proves (i).

Suppose that $\tilde{X}$ is path connected, and let $\tilde{x} \in p^{-1}(\{x_0\})$ be arbitrary. By the definition of path connectedness, there is a path $\tilde{\gamma}$ in $\tilde{X}$ connecting $\tilde{x}_0$ with $\tilde{x}$. Letting $\gamma := p \circ \tilde{\gamma}$ (so that $\tilde{\gamma}$ is trivially the lifting of γ with starting point $\tilde{x}_0$), we obtain that $\phi([\gamma]) = \tilde{x}$. This proves (ii).

For (iii), let $[\gamma_1], [\gamma_2] \in \pi_1(X, x_0)$ be such that $\phi([\gamma_1]) = \phi([\gamma_2])$; that is, the liftings $\tilde{\gamma}_1$ and $\tilde{\gamma}_2$ have the same endpoint, say $\tilde{x}_1$. Consequently, $\tilde{\gamma}_1 \odot \tilde{\gamma}_2^{-1}$ represents an element of $\pi_1\left(\tilde{X}, \tilde{x}_0\right) = \{0\}$, so that there is a path homotopy $\tilde{\Gamma}$ between $\tilde{\gamma}_1 \odot \tilde{\gamma}_2^{-1}$ and $\tilde{x}_0$. Letting $\Gamma := p \circ \tilde{\Gamma}$, we thus obtain a path homotopy between $\gamma_1 \odot \gamma_2^{-1}$ and x_0, so that $[\gamma_1] \cdot [\gamma_2]^{-1} = [\gamma_1 \odot \gamma_2^{-1}] = [x_0]$ and thus $[\gamma_1] = [\gamma_2]$. This proves (iii). $\square$

We now use Theorem 5.2.6 to compute $\pi_1(\mathbb{S}^1, (1,0))$.

Example 5.2.7. Let

$$p \colon \mathbb{R} \to \mathbb{S}^1, \quad x \mapsto (\cos(2\pi x), \sin(2\pi x)),$$

which turns, as we saw in Example 5.2.2, $\mathbb{R}$ into a covering space of $\mathbb{S}^1$, and note that $p^{-1}(\{(1,0)\}) = \mathbb{Z}$. Since $\mathbb{R}$ is simply connected, Theorem 5.2.6 immediately yields that the lifting correspondence $\phi \colon \pi_1(\mathbb{S}^1, (1,0)) \to \mathbb{Z}$ is a bijective map. What remains to be shown is that it is a group homomorphism.

Let $n, m \in \mathbb{Z}$, and let $\gamma_n, \gamma_m \in P(\mathbb{S}^1, (1,0))$ be such that their liftings $\tilde{\gamma}_n, \tilde{\gamma}_m \in P(\mathbb{R}, 0)$ satisfy $\tilde{\gamma}_n(1) = n$ and $\tilde{\gamma}_m(1) = m$. Define

$$\tilde{\alpha} \colon [0,1] \to \mathbb{R}, \quad t \mapsto n + \tilde{\gamma}_m(t),$$

so that $\tilde{\alpha} \in P(\mathbb{R}; n, n + m)$ with $p \circ \tilde{\alpha} = \gamma_m$. The path $\tilde{\gamma}_n \odot \tilde{\alpha}$ is then well defined, and satisfies

$$p \circ (\tilde{\gamma}_n \odot \tilde{\alpha}) = \gamma_n \odot \gamma_m.$$

Since the starting point of $\tilde{\gamma}_n \odot \tilde{\alpha}$ is 0 and its endpoint is $n + m$, we conclude that $\tilde{\gamma}_n \odot \tilde{\alpha}$ is the lifting $\widetilde{\gamma_n \odot \gamma_m}$ of $\gamma_n \odot \gamma_m$ with starting point 0. All in all, we obtain that

$$\phi([\gamma_n] \cdot [\gamma_m]) = \left(\widetilde{\gamma_n \odot \gamma_m} \right)(1) = (\tilde{\gamma}_n \odot \tilde{\alpha})(1) = n + m.$$

This proves the claim.

Exercises

1. Let $n \in \mathbb{N}$, and define
   $$p_n : \mathbb{S}^1 \to \mathbb{S}^1, \quad z \mapsto z^n,$$
   where $\mathbb{S}^1$ is viewed as a subset of $\mathbb{C}$. Show that $(\mathbb{S}^1, p_n)$ is a covering space of $\mathbb{S}^1$.
2. Let $(X, \mathcal{T}_X)$ and $(Y, \mathcal{T}_Y)$ be topological spaces, where Y is discrete, let $X \times Y$ be equipped with the product topology, and let $p : X \times Y \to X$ be the projection onto the first coordinate. Show that $(X \times Y, p)$ is a covering space of X.
3. Let $(X, \mathcal{T})$ be a connected topological space, let $\left((\tilde{X}, \tilde{\mathcal{T}}), p \right)$ be a covering space for $(X, \mathcal{T})$, and suppose that there are $n \in \mathbb{N}$ and $x \in X$ such that $p^{-1}(\{x\})$ has n elements. Show that $p^{-1}(\{y\})$ has n elements for each $y \in X$. (*Hint:* Show that the set $\{y \in X : p^{-1}(\{y\})$ has n elements$\}$ is nonempty and clopen in X.)
4. Let $(X, \mathcal{T})$ be a topological space, and let $\left((\tilde{X}, \tilde{\mathcal{T}}), p \right)$ be a covering space of X such that $\tilde{X}$ is path connected. Show that $p : \tilde{X} \to X$ is a homeomorphism if $\pi_1(X, x_0) = \{0\}$ for some $x_0 \in X$.

Remarks

What we have done in this chapter is to at most dip one toe into the vast ocean algebraic topology really is.

The only fundamental groups we have computed in this chapter are $\{0\}$ and $\mathbb{Z}$, and with the help of Exercise 5.1.5, it is easy to come up with spaces with fundamental group $\mathbb{Z}^n$ for any $n \in \mathbb{N}$. All these examples are abelian. It is *not* true, however, that fundamental groups are generally abelian: the ∞-shaped subspace below of $\mathbb{R}^2$ has the free group in two generators as its fundamental group. This follows from the *Seifert–van Kampen theorem* (see [MASSEY 91] for details).

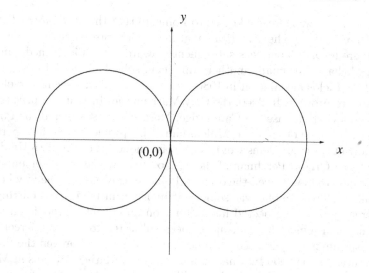

Fig. 5.9: A space with nonabelian fundamental group

The fundamental group is not the only (and by no means the most fundamental!) algebraic invariant of a topological space. As one might guess in view of the symbol $\pi_1(X, x_0)$ for the fundamental group of X at x_0, there are also groups $\pi_n(X, x_0)$ for $n \geq 2$. Unlike the fundamental group, these *higher homotopy groups* are always abelian. Other important invariants studied in algebraic topology are *homology* and *cohomology groups*. Just to define those groups—let alone compute them—requires extensive preparations that go beyond the scope of this book.

Introductions to algebraic topology are [MUNKRES 84] and [MASSEY 91]. The undergraduate textbook [MUNKRES 00] also limits itself to a discussion of the fundamental group and of covering spaces, but covers much more ground than this book.

Even though we treat algebraic topology after set-theoretic topology, it is fair to say that algebraic topology is the older of the two topological disciplines. Attempts to classify various kinds of surfaces go as far back as to the first half of the nineteenth century. The notion of homotopy (along with the definition of the fundamental group) appears for the first time in Poincaré's *Analysis situs* from 1895.

Due also to Poincaré is one of the most famous problems there is in topology—and in all of mathematics—the *Poincaré conjecture*.

Every closed, 3-dimensional manifold that is homotopically equivalent to $\mathbb{S}^3$ is already homeomorphic to $\mathbb{S}^3$.

(A closed, 3-dimensional manifold is a compact topological space that "looks locally like $\mathbb{R}^3$," such as $\mathbb{S}^3$.) One can, of course, generalize the Poincaré conjecture by replacing the number three in it by an arbitrary positive integer n.

For $n = 1, 2$, it was already known to Poincaré that this *generalized Poincaré conjecture* is true. The American mathematician Steven Smale proved the conjecture for $n \geq 5$ and was subsequently awarded the Fields medal in 1966. Then, Michael Freedman, another American, solved the $n = 4$ case (and received the Fields medal for it in 1986). Alas, the $n = 3$ case seemed to elude all attempts to prove it. In 2000, the Clay Mathematics Institute, a private, non-profit organization based in Cambridge, Massachusetts, selected the Poincaré conjecture as one of its seven Millennium Prize problems: the first person to solve any of these problems will receive prize money of one million US dollars.

In 2002, Grigori Perelman of the Steklov Institute in St. Petersburg, Russia, claimed to have solved the original Poincaré conjecture. The way he chose to make his proof public was somewhat unusual: instead of submitting it to a refereed journal, he posted his findings on an e-print server. Even though Perelman's argument has thus never been subjected to formal refereeing, the consensus among experts seems to be that he has indeed proven the Poincaré conjecture. At the 2006 International Congress of Mathematicians in Madrid, Perelman was honored with the Fields medal for this accomplishment; he declined to accept the award. (Needless to say that he hasn't claimed the Millennium Prize money either.)

A

The Classical Mittag-Leffler Theorem Derived from Bourbaki's

This is the Mittag-Leffler theorem from complex variables.

Theorem A.1 (Mittag-Leffler theorem). *Let $\varnothing \neq \Omega \subset \mathbb{C}$ be open, let $\{c_1, c_2, c_3, \ldots\}$ be a discrete subset of Ω, and let $(r_n)_{n=1}^\infty$ be a sequence of rational functions of the form*

$$r_n(z) = \sum_{j=1}^{m_n} \frac{a_{j,n}}{(z - c_n)^j} \qquad (n, m_n \in \mathbb{N}, \, a_{1,n}, \ldots, a_{m_n,n} \in \mathbb{C}, \, z \in \Omega \setminus \{c_n\}).$$

Then there is a meromorphic function f on Ω with $\{c_1, c_2, c_3, \ldots\}$ as its set of poles such that, for each $n \in \mathbb{N}$, the singular part of f at c_n is r_n.

This theorem is usually treated in courses on complex variables, and a proof can be found in probably any text on the subject (such as [CONWAY 78], for example). But what does this theorem have to do with Theorem 2.4.14? Following [ESTERLE 84], we show in this appendix that the Mittag-Leffler theorem can, in fact, be derived from Theorem 2.4.14. Besides Theorem 2.4.14, the proof also requires (of course) some knowledge of complex variables, as well as further topological background from Sections 3.1 to 3.4.

We first need to bring complete metric spaces into the picture. To this end, we prove a lemma.

Lemma A.2. *Let $\varnothing \neq \Omega \subset \mathbb{R}^m$ be open. Then there is a sequence $(K_n)_{n=1}^\infty$ of compact subsets of Ω with the following properties.*

(i) $\Omega = \bigcup_{n=1}^\infty K_n$;

(ii) $K_n \subset \overset{\circ}{K}_{n+1}$ *for all* $n \in \mathbb{N}$;

(iii) *For each* $n \in \mathbb{N}$, *every component of* $\mathbb{R}_\infty^m \setminus K_n$ *contains a component of* $\mathbb{R}_\infty^m \setminus \Omega$.

Proof. If $\Omega = \mathbb{R}^m$, letting $K_n := B_n[0]$ for $n \in \mathbb{N}$ will do.

Hence, we may suppose that $\Omega \neq \mathbb{R}^m$. We may then define

$$K_n := \left\{ x \in \Omega : \|x\| \leq n \text{ and } \mathrm{dist}(x, \mathbb{R}^m \setminus \Omega) \geq \frac{1}{n} \right\} \qquad (n \in \mathbb{N}).$$

It is easy to see that (i) and (ii) are satisfied.

Let $n \in \mathbb{N}$, and let C be a component of $\mathbb{R}_\infty^m \setminus K_n$, where $\mathbb{R}_\infty^m$ is the one-point compactification of $\mathbb{R}^m$.

Case 1: $\infty \in C$.

Let C_∞ be the component of $\mathbb{R}_\infty^m \setminus \Omega$ containing ∞. Then $C_\infty \subset \mathbb{R}_\infty^m \setminus K_n$ is connected and contains ∞. Consequently, $C_\infty \subset C$ must hold.

Case 2: $\infty \notin C$.

The subset $C_0 := \{ x \in \mathbb{R}^m : \|x\| > n \} \cup \{\infty\}$ of $\mathbb{R}_\infty^m \setminus K_n$ is connected. Since C is a component of $\mathbb{R}_\infty^m \setminus K_n$, it follows that either $C_0 \subset C$ or $C_0 \cap C = \varnothing$. Since $\infty \in C_0$ whereas $\infty \notin C$, the first alternative cannot occur, so that $C_0 \cap C = \varnothing$; that is, $\|x\| \leq n$ for all $x \in C$ and therefore, by the definition of K_n, $\mathrm{dist}(x, \mathbb{R}^m \setminus \Omega) < \frac{1}{n}$ for all $x \in C$. Consequently, there is $x_0 \in \mathbb{R}^m \setminus \Omega$ such that $B_{\frac{1}{n}}(x_0) \cap C \neq \varnothing$. Note that $B_{\frac{1}{n}}(x_0) \subset \mathbb{R}^m \setminus K_n$ by the definition of K_n. Since $B_{\frac{1}{n}}(x_0)$ is connected, Proposition 3.4.16 yields that $B_{\frac{1}{n}}(x_0) \subset C$. As in the first case, we see that C contains the component of $\mathbb{R}_\infty^m \setminus \Omega$ containing x_0. $\quad\square$

With the help of Lemma A.2, we can introduce a metric on the space of continuous functions on an open subset of $\mathbb{R}^m$.

Proposition A.3. *Let* $\varnothing \neq \Omega \subset \mathbb{R}^m$ *be open, and let* $(K_n)_{n=1}^\infty$ *be a sequence as in Lemma A.2. For* $f, g \in C(\Omega, \mathbb{F})$ *define*

$$d_n(f, g) := \sup\{ |f(x) - g(x)| : x \in K_n \} \qquad (n \in \mathbb{N})$$

and

$$d(f, g) := \sum_{n=1}^\infty \frac{1}{2^n} \frac{d_n(f, g)}{1 + d_n(f, g)}.$$

Then:

(i) *d is a metric on $C(\Omega, \mathbb{F})$ such that $d(f+h, g+h) = d(f, g)$ for all $f, g, h \in C(\Omega, \mathbb{F})$;*

(ii) *The topology on $C(\Omega, \mathbb{F})$ induced by d is $\mathcal{T}_{\mathcal{K}}|_{C(\Omega, \mathbb{F})}$, where $\mathcal{K}$ is the collection of all compact subsets of Ω;*

(iii) *The metric space $(C(\Omega, \mathbb{F}), d)$ is complete.*

Proof. (i) is clear.

For (ii), let $\mathcal{C} := \{ K_n : n \in \mathbb{N} \}$. It is routine to check that d induces the topology $\mathcal{T}_{\mathcal{C}}|_{C(\Omega, \mathbb{F})}$. Since $\mathcal{C} \subset \mathcal{K}$, it is clear that $\mathcal{T}_{\mathcal{K}}|_{C(\Omega, \mathbb{F})}$ is finer than $\mathcal{T}_{\mathcal{C}}|_{C(\Omega, \mathbb{F})}$. On the other hand, $\left\{ \mathring{K}_n : n \in \mathbb{N} \right\}$ is an open cover for Ω. Hence, for any $K \in \mathcal{K}$, there is $n \in \mathbb{N}$ such that $K \subset \mathring{K}_n \subset K_n$. It follows that $\mathcal{T}_{\mathcal{K}}|_{C(\Omega, \mathbb{F})}$ and $\mathcal{T}_{\mathcal{C}}|_{C(\Omega, \mathbb{F})}$ coincide.

Let $(f_n)_{n=1}^\infty$ be a Cauchy sequence in $(C(\Omega,\mathbb{F}),d)$. Then, for each $x \in \Omega$, the sequence $(f_n(x))_{n=1}^\infty$ is a Cauchy sequence in $\mathbb{F}$, so that $f(x) := \lim_{n\to\infty} f_n(x)$ exists. It is routine to check that $(f_n)_{n=1}^\infty$ converges to $f \colon \Omega \to \mathbb{F}$ uniformly on each $K \in \mathcal{K}$. Let $x_0 \in \Omega$, and let $\epsilon > 0$ be such that $\overline{B_\epsilon(x_0)} \subset \Omega$. Since $(f_n)_{n=1}^\infty$ converges to f uniformly on $\overline{B_\epsilon(x_0)}$, it follows that $f|_{\overline{B_\epsilon(x_0)}}$ is continuous, and since $\overline{B_\epsilon(x_0)}$ is a neighborhood of x_0, the function f is continuous at x_0. This proves (iii). $\square$

To prove Theorem A.1, we apply Theorem 2.4.14 not to all of $C(\Omega,\mathbb{C})$, but to a subspace.

Definition A.4. *Let $\varnothing \neq \Omega \subset \mathbb{C}$ be open. Then $H(\Omega)$ denotes the space of all holomorphic functions on Ω.*

For the following corollary of Proposition A.3, we identify $\mathbb{C}$ with $\mathbb{R}^2$.

Corollary A.5. *Let $\varnothing \neq \Omega \subset \mathbb{C}$ be open, and let d be as in Proposition A.3. Then $H(\Omega)$ is a closed subspace of $(C(\Omega,\mathbb{C}),d)$ (and therefore complete).*

Proof. Let $(f_n)_{n=1}^\infty$ be a sequence in $H(\Omega)$ that converges to $f \in C(\Omega,\mathbb{C})$ with respect to d and thus, by Proposition A.3(ii), uniformly on all compact subsets of Ω. It is well known that this forces f to be holomorphic, too (see, for example, [CONWAY 78, 2.1 Theorem]). $\square$

We can now prove Theorem A.1 with the help of Theorem 2.4.14.

Proof (of Theorem A.1). Let $(K_n)_{n=1}^\infty$ be a sequence as specified by Lemma A.2, and let $\Omega_{n-1} := \overset{\circ}{K}_n$ for $n \in \mathbb{N}$. For each $n \in \mathbb{N}_0$, let $\tilde{d}_n$ be a metric on $H(\Omega_n) \subset C(\Omega_n,\mathbb{C})$ as specified by Proposition A.3.

Let $n \in \mathbb{N}_0$ be fixed, and let $S_n := \{m \in \mathbb{N} : c_m \in \Omega_n\}$. Since K_{n+1} is compact, and since $\{c_1,c_2,\ldots\}$ is discrete, each S_n is finite (and possibly empty). Hence, the rational function $R_n := \sum_{m \in S_n} r_m$ is well defined (the sum is finite). Let X_n be the set of those meromorphic functions f on Ω_n such that $f - R_n$ has a holomorphic extension to all of Ω_n, and define a metric on it via

$$d_n(f,g) := \tilde{d}_n(f - R_n, g - R_n) \qquad (f,g \in X_n).$$

It follows from Corollary A.5 that (X_n,d_n) is a complete metric space. For $n \in \mathbb{N}$, let $\phi_n \colon X_n \to X_{n-1}$ denote the restriction map. In view of Proposition A.3(ii), it is clear that ϕ_n is continuous.

We claim that ϕ_n has dense range. Let $g \in X_{n-1}$, so that $g - R_{n-1} \in H(\Omega_{n-1})$. Since the rational functions r_m for $m \in S_n \setminus S_{n-1}$ have their poles off Ω_{n-1}, it follows that $g - R_n$ is in $H(\Omega_{n-1})$ as well. Due to Lemma A.2 and Runge's approximation theorem [CONWAY 78, 1.14 Corollary], we can find a sequence $(q_m)_{m=1}^\infty$ of rational functions with poles off Ω_{n-1} (which therefore belong to $H(\Omega_{n-1})$) such that $\tilde{d}_{n-1}(g - R_n, q_m) \to 0$. It follows that

$$d_{n-1}(g, \phi_n(q_m + R_n)) = \tilde{d}_{n-1}(g - R_{n-1}, q_m + R_n - R_{n-1})$$
$$= \tilde{d}_{n-1}(g - R_n + \underbrace{(R_n - R_{n-1})}_{\in H(\Omega_{n-1})}, q_m + \underbrace{(R_n - R_{n-1})}_{\in H(\Omega_{n-1})})$$
$$= \tilde{d}_{n-1}(g - R_n, q_m)$$
$$\to 0$$

as $m \to \infty$, and consequently, $\phi_n(X_n)$ is dense in X_{n-1}.

From Theorem 2.4.14, we conclude that $\bigcap_{n=1}^{\infty}(\phi_1 \circ \cdots \circ \phi_n)(X_n)$ is dense in X_0 and thus, in particular, is not empty. Let $(g_n)_{n=0}^{\infty}$ be a sequence such that $g_n \in X_n$ for $n \in \mathbb{N}_0$ and $\phi_n(g_n) = g_{n-1}$ for $n \in \mathbb{N}$. Define $f \colon \Omega \setminus \{c_1, c_2, \ldots\} \to \mathbb{C}$ by letting $f(z) := g_n(z)$ if $z \in \Omega_n \setminus \{c_m : m \in S_n\}$. Since $\Omega = \bigcup_{n=1}^{\infty} \Omega_n$, this defines a meromorphic function on Ω with the required properties. $\square$

B

Failure of the Heine–Borel Theorem in Infinite-Dimensional Spaces

We first show that the Heine–Borel theorem holds in all finite-dimensional, normed spaces.

The following is the crucial assertion for this.

Proposition B.1. *Let E be a finite-dimensional, linear space (over $\mathbb{F} = \mathbb{R}$ or $\mathbb{F} = \mathbb{C}$), and let $\|\cdot\|$ and $\|\|\cdot\|\|$ be norms on E. Then there is a constant $C \geq 0$ such that*

$$\|x\| \leq C\|\|x\|\| \quad and \quad \|\|x\|\| \leq C\|x\| \qquad (x \in E).$$

Proof. Let $e_1, \ldots, e_n \in E$ be a basis for E. For $x = \lambda_1 e_1 + \cdots + \lambda_n e_n$, let

$$|x| := \max\{|\lambda_1|, \ldots, |\lambda_n|\}.$$

Clearly, $|\cdot|$ is a norm on E.

Set $C_1 := \|e_1\| + \cdots + \|e_n\|$, and note that

$$\|x\| \leq |\lambda_1|\|e_1\| + \cdots + |\lambda_n|\|e_n\| \leq C_1|x| \qquad (x \in E).$$

Next, we show that there is $C_2 \geq 0$ with $|x| \leq C_2\|x\|$ for all $x \in E$.

Assume otherwise. Then there is a sequence $(x_m)_{m=1}^{\infty}$ in E with $|x_m| > m\|x_m\|$ for $m \in \mathbb{N}$. Let

$$y_m := \frac{x_m}{|x_m|} \qquad (m \in \mathbb{N}).$$

For each $m \in \mathbb{N}$, there are unique $\lambda_{1,m}, \ldots, \lambda_{n,m} \in \mathbb{F}$ with $y_m = \sum_{j=1}^{n} \lambda_{j,m} e_j$. It follows that

$$1 = |y_m| = \max\{|\lambda_{1,m}|, \ldots, |\lambda_{n,m}|\} \qquad (m \in \mathbb{N}).$$

In particular, the sequence $((\lambda_{1,m}, \ldots, \lambda_{n,m}))_{m=1}^{\infty}$ is bounded in $\mathbb{F}^n$ and thus has—by the Bolzano–Weierstraß theorem (for $\mathbb{R}^n$ if $\mathbb{F} = \mathbb{R}$ and for $\mathbb{R}^{2n}$ if $\mathbb{F} = \mathbb{C}$)—a convergent subsequence, say $((\lambda_{1,m_k}, \ldots, \lambda_{n,m_k}))_{k=1}^{\infty}$ with limit

$(\lambda_1, \ldots, \lambda_n)$. It follows that $(y_{m_k})_{k=1}^{\infty}$ converges, with respect to $|\cdot|$, to $y :=$ $\lambda_1 e_1 + \cdots + \lambda_n e_n$, so that necessarily $|y| = 1$ and thus $y \neq 0$. Since $\|\cdot\| \leq C_1|\cdot|$, we see that $y = \lim_{k \to \infty} y_{m_k}$ as well with respect to $\|\cdot\|$. However,

$$\|y_m\| = \left\|\frac{x_m}{|x_m|}\right\| = \frac{\|x_m\|}{|x_m|} < \frac{1}{m} \to 0,$$

so that $y = 0$. This is impossible.

For $C' := \max\{C_1, C_2\}$, we have

$$\|x\| \leq C'|x| \quad \text{and} \quad |x| \leq C'\|x\| \qquad (x \in E),$$

and in a similar vein, we obtain $C'' \geq 0$ such that

$$|||x||| \leq C''|x| \quad \text{and} \quad |x| \leq C'''|||x||| \qquad (x \in E).$$

Consequently, with $C := C'C''$,

$$\|x\| \leq C|||x||| \quad \text{and} \quad |||x||| \leq C\|x\| \qquad (x \in E)$$

holds. □

As an immediate consequence, any two norms on a finite-dimensional vector space E yield equivalent metrics, and if E is a Banach space with respect to one norm, it is a Banach space with respect to *every* norm. Hence, if $\dim E = n$ and if $e_1, \ldots, e_n$ is a basis of E, the map

$$\mathbb{F}^n \to E, \quad (\lambda_1, \ldots, \lambda_n) \mapsto \lambda_1 e_1 + \cdots + \lambda_n e_n$$

is continuous with continuous inverse and carries Cauchy sequences to Cauchy sequences (as does its inverse).

We therefore obtain the following.

Corollary B.2. *Let E be a finite-dimensional, normed space. Then E is a Banach space, and a subset of E is compact if and only if it is closed and bounded.*

Combining this with Proposition 2.4.5(ii) yields the following.

Corollary B.3. *Let E be a normed space, and let F be a finite-dimensional subspace of E. Then F is closed in E.*

By Corollary B.2, the Heine–Borel theorem holds true in any finite-dimensional normed space. For the converse, we require the following.

Lemma B.4 (Riesz' lemma). *Let E be a normed space, and let F be a closed, proper (i.e., $F \neq E$), subspace of E. Then, for each $\theta \in (0, 1)$, there is $x_\theta \in E$ with $\|x_\theta\| = 1$, and $\|x - x_\theta\| \geq \theta$ for all $x \in F$.*

Proof. Let $x_0 \in E \setminus F$, and let $\delta := \mathrm{dist}(x_0, F)$. If $\delta = 0$, the closedness of F implies $x_0 \in F$, which is a contradiction. Hence, $\delta > 0$ must hold. Since $\theta \in (0, 1)$, we have $\delta < \frac{\delta}{\theta}$. Choose $y_\theta \in F$ with $0 < \|x_0 - y_\theta\| < \frac{\delta}{\theta}$, and let

$$x_\theta := \frac{y_\theta - x_0}{\|y_\theta - x_0\|},$$

so that trivially $\|x_\theta\| = 1$. Let $x \in F$, and note that

$$\|x - x_\theta\| = \left\| x - \frac{y_\theta - x_0}{\|y_\theta - x_0\|} \right\| = \frac{1}{\|y_\theta - x_0\|} \|\|y_\theta - x_0\|x - y_\theta + x_0\|.$$

Since $x, y_\theta \in F$, we have $\|y_\theta - x_0\|x - y_\theta \in F$ as well, so that

$$\|\|y_\theta - x_0\|x - y_\theta + x_0\| \geq \mathrm{dist}(x_0, F) = \delta.$$

Eventually, we obtain

$$\|x - x_\theta\| = \frac{1}{\|y_\theta - x_0\|} \|\|y_\theta - x_0\|x - y_\theta + x_0\| > \frac{\theta}{\delta}\delta = \theta.$$

Since $x \in F$ was arbitrary, this completes the proof. □

We can now prove the following.

Theorem B.5. *For a normed space E, the following are equivalent.*

(i)   *Every closed and bounded subset of E is compact.*
(ii) *The closed unit sphere of E is compact.*
(iii) $\dim E < \infty$.

Proof. (i) $\Longrightarrow$ (ii) is trivial.

(ii) $\Longrightarrow$ (iii): Suppose that $\dim E = \infty$. We construct a sequence in $S_1[0]$ that has no convergent subsequence, so that $S_1[0]$ cannot be compact by Theorem 2.5.10

Choose $x_1 \in E$ with $\|x_1\| = 1$. Since $\dim E = \infty$, the one-dimensional space F_1 spanned by x_1 is not all of E. By Riesz' lemma, there is thus $x_2 \in E$ such that $\|x_2 - x\| \geq \frac{1}{2}$ for $x \in F_1$, so that, in particular, $\|x_2 - x_1\| \geq \frac{1}{2}$. Since $\dim E = \infty$, the two-dimensional space F_2 spanned by $\{x_1, x_2\}$ is also not all of E. Again by Riesz' lemma, there is thus $x_3 \in E$ such that $\|x_3 - x\| \geq \frac{1}{2}$ for $x \in F_2$, and thus, in particular, $\|x_3 - x_j\| \geq \frac{1}{2}$ for $j = 1, 2$. Let F_3 be the linear span of $\{x_1, x_2, x_3\}$, so that $F_3 \neq E$. Appealing again to Riesz' lemma, we obtain $x_4 \in E$, and so on.

Inductively, we thus obtain a sequence $(x_n)_{n=1}^\infty$ in $S_1[0]$ such that

$$\|x_n - x_m\| \geq \frac{1}{2} \qquad (n \neq m).$$

It is clear that no subsequence of $(x_n)_{n=1}^\infty$ can be a Cauchy sequence.

Finally, (iii) $\Longrightarrow$ (i) is Corollary B.2. □

The Arzelà–Ascoli Theorem

As we have seen in Example 2.5.13, the Heine–Borel theorem is false for $C([0,1], \mathbb{F})$ (and, more generally, for *every* infinite-dimensional normed space; see Appendix B).

The Arzelà–Ascoli theorem can be thought of as the right substitute for the Heine–Borel theorem in spaces of continuous functions. In this appendix, we derive it from Tychonoff's theorem.

For the statement of the Arzelà–Ascoli theorem, we require two notions: that of relative compactness, which was introduced in Exercise 2.5.7, and that of equicontinuity.

Definition C.1. *Let $(X, \mathcal{T})$ be a topological space, and let (Y, d) be a metric space. Then a family $\mathfrak{F}$ of functions from X to Y is said to be* equicontinuous *at $x_0 \in X$ if, for each $\epsilon > 0$, there is $N \in \mathcal{N}_{x_0}$ such that $d(f(x_0), f(x)) < \epsilon$ for all $f \in \mathfrak{F}$ and $x \in N$. If $\mathfrak{F}$ is equicontinuous at every point of X, we call $\mathfrak{F}$* equicontinuous.

If $\mathfrak{F}$ consists only of one function, say f, then $\mathfrak{F}$ is equicontinuous if and only if f is continuous.

Let $(K, \mathcal{T})$ be a compact topological space, let (Y, d) be a metric space, and let $f \colon K \to Y$ be continuous. Then $f(K)$ is compact and therefore has finite diameter, which means that f is actually in $C_b(K, Y)$. In the following result, we have $C(K, Y) = C_b(K, Y)$ equipped with the metric D introduced in Example 2.1.2(d).

Theorem C.2 (Arzelà–Ascoli theorem). *Let $(K, \mathcal{T})$ be a compact topological space, and let (Y, d) be a complete metric space. Then the following are equivalent for $\mathfrak{F} \subset C(K, Y)$.*

(i) $\mathfrak{F}$ *is relatively compact in $C(K, Y)$.*
(ii) (a) $\{f(x) : f \in \mathfrak{F}\}$ *is relatively compact in Y for each $x \in X$, and*
  (b) $\mathfrak{F}$ *is equicontinuous.*

Proof. (i) $\implies$ (ii): For $x \in K$, let

$$\pi_x \colon C(K,Y) \to Y, \quad f \mapsto f(x).$$

Then π_x is continuous, so that $\pi_x\left(\overline{\mathfrak{F}}\right)$ is compact in Y and contains $\{f(x) : f \in \mathfrak{F}\}$. Consequently, $\{f(x) : f \in \mathfrak{F}\}$ is relatively compact in Y. This proves (ii)(a).

Assume towards a contradiction that (ii)(b) is false; that is, there are $x_0 \in X$ and $\epsilon_0 > 0$ such that, for each $N \in \mathcal{N}_{x_0}$, there are $f_N \in \mathfrak{F}$ and $x_N \in N$ such that $d(f_N(x_0), f_N(x_N)) \geq \epsilon_0$. Since $\overline{\mathfrak{F}}$ is compact, the net $(f_N)_{N \in \mathcal{N}_{x_0}}$, where $\mathcal{N}_{x_0}$ is ordered by reversed set inclusion, has a subnet $(f_\alpha)_{\alpha \in \mathbb{A}}$ converging (with respect to D) to some $f \in \overline{\mathfrak{F}}$. Let $N_0 \in \mathcal{N}_{x_0}$ be such that $d(f(x_0), f(x)) < \frac{\epsilon_0}{3}$ for $x \in N_0$ (this is possible because f is continuous), let $\phi \colon \mathbb{A} \to \mathcal{N}_{x_0}$ be the cofinal map associated with the subnet $(f_\alpha)_{\alpha \in \mathbb{A}}$, and let $\alpha \in \mathbb{A}$ be such that $D(f_\alpha, f) < \frac{\epsilon_0}{3}$ and $\phi(\alpha) \subset N_0$. We then have:

$$
\begin{aligned}
d(f_\alpha&(x_0), f_\alpha(x_{\phi(\alpha)})) \\
&\leq d(f_\alpha(x_0), f(x_0)) + d(f(x_0), f(x_{\phi(\alpha)})) + d(f(x_{\phi(\alpha)}), f_\alpha(x_{\phi(\alpha)})) \\
&\leq D(f_\alpha, f) + d(f(x_0), f(x_{\phi(\alpha)})) + D(f_\alpha, f) \\
&< \frac{2\epsilon_0}{3} + d(f(x_0), f(x_{\phi(\alpha)})) \\
&< \frac{2\epsilon_0}{3} + \frac{\epsilon_0}{3}, \qquad \text{because } \phi(\alpha) \subset N_0, \\
&= \epsilon_0.
\end{aligned}
$$

This contradicts the choices of f_N and x_N for $N \in \mathcal{N}_{x_0}$. (This part of the proof has not made any reference to the completeness of Y or to the compactness of K.)

(ii) $\Longrightarrow$ (i): Since (a) and (b) are not affected if we replace $\mathfrak{F}$ by its closure, we can suppose without loss of generality that $\mathfrak{F}$ is closed.

Let $(f_\alpha)_\alpha$ be a net in $\mathfrak{F}$. We show that it has a convergent subnet.

For $x \in K$, let $K_x := \overline{\{f(x) : f \in \mathfrak{F}\}}$, so that K_x is compact by (a). Tychonoff's theorem then yields the compactness of the topological product $\prod_{x \in K} K_x$. Hence, $(f_\alpha)_\alpha$ has a subnet $(f_\beta)_{\beta \in \mathbb{B}}$ such that $(f_\beta(x))_{\beta \in \mathbb{B}}$ converges for each $x \in K$. By Exercise 3.2.12(a), this means in particular that, for each $\epsilon > 0$ and $x \in K$, there is $\beta_{x,\epsilon} \in \mathbb{B}$ such that $d(f_\beta(x), f_\gamma(x)) < \epsilon$ for all $\beta, \gamma \in \mathbb{B}$ with $\beta_{x,\epsilon} \preceq \beta, \gamma$.

Fix $\epsilon > 0$. For each $x \in X$, choose an open neighborhood U_x of x such that $d(f(x), f(x')) < \frac{\epsilon}{3}$ for $x' \in U_x$. Clearly, $\{U_x : x \in K\}$ is an open cover for K. Since K is compact, there are $x_1, \ldots, x_n \in K$ such that

$$K = U_{x_1} \cup \cdots \cup U_{x_n}.$$

Choose $\beta_\epsilon \in \mathbb{B}$ such that $d(f_\beta(x_j), f_\gamma(x_j)) < \frac{\epsilon}{3}$ for all $j = 1, \ldots, n$ and $\beta, \gamma \in \mathbb{B}$ with $\beta_\epsilon \preceq \beta, \gamma$. Let $x \in K$, and choose $j \in \{1, \ldots, n\}$ such that $x \in U_{x_j}$. Then we have for $\beta, \gamma \in \mathbb{B}$ with $\beta_\epsilon \preceq \beta, \gamma$:

$$d(f_\beta(x), f_\gamma(x)) \leq d(f_\beta(x), f_\beta(x_j)) + d(f_\beta(x_j), f_\gamma(x_j)) + d(f_\gamma(x_j), f_\gamma(x))$$
$$< \frac{\epsilon}{3} + \frac{\epsilon}{3} + \frac{\epsilon}{3}$$
$$= \epsilon.$$

It follows that $D(f_\beta, f_\gamma) \leq \epsilon$ for $\beta, \gamma \in \mathbb{B}$ with $\beta_\epsilon \preceq \beta, \gamma$, so that $(f_\beta)_{\beta \in \mathbb{B}}$ is a Cauchy net in $C(K, Y)$. Since $B(K, Y)$ is complete by Example 2.4.4(c), it follows from Exercise 3.2.12(b), that $(f_\beta)_{\beta \in \mathbb{B}}$ converges to some $f \in B(K, Y)$. As in Example 2.4.6, where the case of the domain being a metric space was treated, one sees that $f \in C(K, Y)$. $\square$

Let $(K, \mathcal{T})$ be a compact topological space. Then $C(K, \mathbb{F})$ is a normed space, so that it makes sense to speak of bounded sets. As an immediate consequence of Theorem C.2, we obtain what may be construed as an infinite-dimensional Heine–Borel theorem.

Corollary C.3. *Let* $(K, \mathcal{T})$ *be a compact topological space. Then a subset of* $C(K, \mathbb{F})$ *is compact if and only if it is closed, bounded, and equicontinuous.*

References

[ALEXANDROFF & HOPF 35] PAUL (=PAVEL) ALEXANDROFF and HEINZ HOPF. 1935. *Topologie*, Band I. Berlin: Springer Verlag.

[BOURBAKI 60] NICOLAS BOURBAKI. 1960. *Topologie générale*, Chapître II. Paris: Hermann.

[CHERNOFF 92] PAUL R. CHERNOFF. 1992. A simple proof of Tychonoff's theorem via nets. *American Mathematical Monthly* 99, 932–934.

[CONWAY 78] JOHN B. CONWAY. 1978. *Functions of One Complex Variable*. 2nd ed. New York: Springer Verlag.

[DALES 78] H. GARTH DALES. 1978. Automatic continuity: A survey. *Bulletin of the London Mathematical Society* 10, 129–183.

[ESTERLE 84] JEAN ESTERLE. 1984. Mittag-Leffler methods in the theory of Banach algebras and a new approach to Michael's problem. In *Proceedings of the Conference on Banach Algebras and Several Complex Variables* (New Haven, 1983). Contemporary Mathematics 32, 107–129. Providence, RI: American Mathematical Society.

[FARENICK 01] DOUGLAS R. FARENICK. 2001. *Algebras of Linear Transformations*. New York: Springer-Verlag.

[FRÉCHET 06] MAURICE FRÉCHET. 1906. Sur quelques points du calcul fonctionnel. *Rendiconti del Circolo Matematico di Palermo* XXII, 1–74.

[HALMOS 74] PAUL R. HALMOS. 1974. *Naive Set Theory*. New York: Springer-Verlag.

[HAUSDORFF 14] FELIX HAUSDORFF. 1914. *Grundzüge der Mengenlehre*. Leipzig: Verlag von Veit.

[JAMESON 74] GRAHAM J. O. JAMESON. 1974. *Topology and Normed Spaces*. Londong: Chapman & Hall, London.

[KELLEY 50] JOHN L. KELLEY. 1950. The Tychonoff product theorem implies the axiom of choice. *Fundamenta Mathematica* 37, 75–76.

[KELLEY 55] JOHN L. KELLEY. 1955. *General Topology*. New York: Van Nostrand.

[MACHADO 77] SILVIO MACHADO. 1977. On Bishop's generalization of the Weierstrass–Stone theorem. *Indagationes Mathematicae* 39, 218–224.

[MASSEY 91] WILLIAM S. MASSEY. 1991. *A Basic Course in Algebraic Topology*. New York: Springer Verlag.

[MUNKRES 84] JAMES R. MUNKRES. 1984. *Elements of Algebraic Topology*. Reading, MA: Addison-Wesley.

[MUNKRES 00] JAMES R. MUNKRES. 2000. *Topology.* 2nd ed. Upper Saddle River: Prentice-Hall.

[MURPHY 90] GERARD J. MURPHY. 1990. *C*-Algebras and Operator Theory.* Boston: Academic Press.

[RANSFORD 84] THOMAS J. RANSFORD. 1984. A short elementary proof of the Bishop–Stone–Weierstrass theorem. *Mathematical Proceedingds of the Cambridge Philosophical Society* 96, 309–311.

[SIMMONS 63] GEORGE F. SIMMONS. 1963. *Introduction to Topology and Modern Analysis.* International Student Edition. Singapore: McGraw-Hill.

[STONE 37] MARSHALL H. STONE. 1937. Applications of the theory of Boolean rings to general topology. *Transactions of the American Mathematical Society* 41, 375–481.

[WILLARD 70] STEPHEN WILLARD. 1970. *General Topology.* Reading, MA: Addison-Wesley.

Index

A-antisymmetric set, 121
accumulation point, 82
  partial, 84
Alaoglu–Bourbaki theorem, 88
Alexandroff, Pavel S., 107, 129
Alexandrov, Pavel S., *see* Alexandroff,
  Pavel S.
algebra, 121
  unital, 121
Analysis situs, 107, 155
Arzelà–Ascoli theorem, 85, 165
axiom of choice, 20

Baire's
  category theorem, 59
  theorem, 48
ball
  closed, 30
  open, 28
Banach space, 41
Banach's fixed point theorem, 51
base
  for a neighborhood system, 65
  for a topology, 69
Bernstein polynomial, 128
bijection, *see* function, bijective
Bing–Nagata–Smirnoff theorem, *see*
  Nagata–Smirnoff theorem
Bolzano–Weierstraß theorem, 161
boundary, 33, 70
Bourbaki
  Charles Denis, 59
  Nicolas, 59
Bourbaki's Mittag-Leffler theorem,
  47
Brouwer's fixed point theorem, 144
  for $n = 1, 2$, 143

C^*-algebra, 130
Cantor set, 95
Cantor's intersection theorem, 44

Cantor, Georg, 21
Cantor–Bernstein theorem, 15
cardinal, 16
  finite, 16
  infinite, 16
cardinal number, *see* cardinal
cardinality, 7
  less than or equal to, 15
  the same, 13
Cartesian product, 9, 17, 18, 26
Cauchy
  net, 79
  sequence, 41
Čech, Eduard, 130
Chernoff, Paul R., 107
choice function, 18
clopen set, 91
closed
  ball, 30
  interval, 6
  manifold, 155
  path, 138
    base point of, 138
  set, 30, 63
closure, 30, 66
coffee cup, *see* doughnut
Cohen, Paul, 22
compact set, 52, 79
compactification
  one-point, 86
  Stone–Čech, 118
compactness, 52, 79
comparable topologies, 73
complement, 8
completely regular space, 101
completeness, 41
completion, 44, 46, 56
component, 93
composition, 11
concatenation of paths, 98

connectedness, 91
continuity, 37, 72
  at a point, 36, 72
continuum hypothesis, 22
convergence
  coordinatewise, 40, 83
  of a net, 74
  of a sequence, 35, 72
  pointwise, 51, 58, 75
  uniform, 58, 75
convex set, 89
coordinate, 9, 17
coordinate projection, 10, 83
coordinatewise convergence, 40, 83
covering
  map, 149
  space, 149

de Morgan's rules, 12
dense subset, 31, 68
diameter, 44
Dini's lemma, 88
directed set, 73
disjoint sets, 8
distance, 34
  Euclidean, 23
domain, 9
doughnut, *see* coffee cup

element, 5
  maximal, 19
empty set, 5
equicontinuity, 165
  at a point, 165
equivalence
  class, 20
  relation, 20, 138
Esterle, Jean, 59

finite intersection property, 79
Fréchet, Maurice, 59, 107
Freedman, Michael, 156
French railroad metric, 25
function, 9
  bijective, 11

bounded, 24
continuous, 36, 37, 72
  nowhere differentiable, 49
  vanishing at infinity, 126
holomorphic, 159
injective, 11
inverse, 12
isometric, 45
meromorphic, 157
  singular part of, 157
rational, 157
Riemann integrable, 74
surjective, 11
uniformly continuous, 58
fundamental group, 138
  of $\mathbb{S}^1$, 153

Gelfand–Naimark theorem, 130
group
  cohomology, 155
  fundamental, 138
  of $\mathbb{S}^1$, 153
  higher homotopy, 155
  homology, 155
  homomorphism, 142
  isomorphism, 142, 144, 146
  topological, 99, 148

half-open interval, 6
Hamel basis, 20
Hausdorff space, 62
Hausdorff, Felix, 59, 107, 108
Heine–Borel theorem, 56
  failure of, 161
Hilbert, David, 21
homeomorphic, 80, 155
homeomorphism, 80
homotopic, 134
homotopically equivalent, 135, 155
homotopy, 134
  equivalence, 135, 146
homotopy type, *see* homotopically
  equivalent

ideal

maximal, 20, 116
  prime, 63
idempotent, 120
identity map, 10
image, 10
  inverse, 10
index, 8
  set, 8
infinitude of primes, 71
injection, *see* function, injective
interior, 34, 70
intermediate value theorem, 92
intersection, 8
interval
  closed, 6
  degenerate, 6
  half-open, 6
  open, 6
inverse
  function, 12
  image, 10
involution, 130
isometry, 45

Jameson, Graham J. O., 107
Jordan content, 95

Kelley, John L., 107
Kolmogorov space, *see* T_0-space
Kuratowski closure operation, 67
Kuratowski, Kazimierz, 107

Lebesgue number, 58
Lebesgue's covering lemma, 58
lifting
  correspondence, 153
  of a path, 150
  of a path homotopy, 151
limit
  of a net, 74
  of a sequence, 35
  uniqueness
    in Hausdorff spaces, 76
    in metric spaces, 35
linear

functional, 88
  space, 24
    finite-dimensional, 161
loop, *see* closed path

Machado, Silvio, 130
manifold, 155
map, *see* function
  cofinal, 81
  identity, 10
  quotient, 78
maximal
  element, 19
  ideal, 20, 116
metric, 24
  French railroad, 25
metric space, 24
  complete, 41, 54
  completion of, 44
  discrete, 27
  separable, 31, 54
  sequentially compact, 55
  subspace of, 24
  totally bounded, 55
Mittag-Leffler theorem, 157
  Bourbaki's, 47
Munkres, James R., 107

Nagata–Smirnoff theorem, 130
neighborhood, 29, 64
  basic, 65
net, 73
  Cauchy, 79
  convergent, 74
noncommutative topology, 130
norm, 24
normal space, 103
normed space, 24
number
  algebraic, 17
  transcendental, 17

one-point compactification, 86
open
  ball, 28

cover, 52, 79
interval, 6
set, 28, 61
ordered
    n-tuple, 17
    pair, 9
    set, 19
ordering, 18

partition, 74
path, 89
    closed, 138
        base point of, 138
    connecting two points, 89
    endpoint of, 138
    homotopic, 136
    homotopy, 136
        lifting of, 151
    lifting of, 150
    reversed, 98
    starting point of, 138
path connectedness, 89
Perelman, Grigori, 156
Poincaré conjecture, 155
    generalized, 156
Poincaré, Henri, 107, 155
pole, 157
positive definiteness, 24
power set, 7
prime
    ideal, 63
    number, 63, 71
product
    Cartesian, 9, 17, 18, 26
    topological, 83
    topology, 83
pseudometric, 59

quotient
    map, 78
    space, 71
    topology, 71

range, 10
Ransford, Thomas J., 130

regular space, 108
relation
    equivalence, 20, 138
    reflexive, 20
    symmetric, 20
    transitive, 20
restriction, 10
retract, 142
retraction, 142
Riemann
    sphere, 86
    sum, 74
Riesz' lemma, 162
ring, 63, 116
    commutative, 20
    homomorphism, 118
Runge's approximation theorem, 159
Russell's antinomy, 7
Russell, Bertrand, 21

Seifert–van Kampen theorem, 154
semimetric, 26
seminorm, 59, 122
sequence, 10
    Cauchy, 41
    convergent, 35, 72
    generalized, see net
set, 5
    A-antisymmetric, 121
    clopen, 91
    closed, 30, 63
    compact, 52, 79
    convex, 89
    countable, 14
    countably infinite, 14
    directed, 73
    empty, 5
    finite, 7
    of all sets, 8
    open, 28, 61
    ordered, 19
        totally, 19
    relatively compact, 58, 165
    star-shaped, 148
    uncountable, 14

set-theoretic difference, 8
sheet, 149
Simmons, George F., 107
singleton, 7
Smale, Steven, 156
Sorgenfrey
    line, 102
    plane, 105
    topology, 102
space
    Banach, 41
    Hausdorff, 62
    linear, 24
        finite-dimensional, 161
    metric, 24
        complete, 41, 54
        discrete, 27
        separable, 31, 54
        sequentially compact, 55
        totally bounded, 55
    normed, 24
    quotient, 71
    topological, 61
        chaotic, 62
        completely regular, 101
        connected, 91
        disconnected, 91
        discrete, 62
        first countable, 66
        Hausdorff, 62
        Lindelöf, 106
        locally (path) connected, 96
        locally compact, 85
        metrizable, 62, 112, 126
        normal, 103
        path connected, 89
        regular, 108
        σ-compact, 106
        second countable, 71, 112
        separable, 69
        simply connected, 153
        T_0, 100
        T_1, 100
        totally disconnected, 96
        zero-dimensional, 99

Stone, Marshall H., 130
Stone–Čech compactification, 118
Stone–Weierstraß theorem
    complex, 124
    for locally compact spaces, 127
    real, 125
subalgebra, 121
    unital, 121
subbase, 69
subcover, 52
subnet, 81
    convergent, 82, 85
subsequence, 10
    convergent, 54
subset, 6
    dense, 31, 68
    nowhere dense, 59
    of the first category, 59
    of the second category, 59
    proper, 6
subspace
    of a metric space, 24
    of a topological space, 62
surjection, see function, surjective
symmetry, 24

T_0-space, 100
T_1-space, 100
T_2-space, see Hausdorff space
$T_{3\frac{1}{2}}$-space, see completely regular
        space
Tietze's extension theorem, 113
Tikhonov, Andrei N., see Tychonoff,
        Andrey N.
topological space, 61
    chaotic, 62
    completely regular, 101
    connected, 91
    disconnected, 91
    discrete, 62
    first countable, 66
    Hausdorff, 62
    Lindelöf, 106
    locally (path) connected, 96
    locally compact, 85

metrizable, 62, 112, 126
normal, 103
path connected, 89
regular, 108
σ-compact, 106
second countable, 71, 112
separable, 69
simply connected, 153
T_0, 100
T_1, 100
totally disconnected, 96
zero-dimensional, 99
topology, 61
box, 88
coarser, 73
finer, 73
of coordinatewise convergence, 83
of pointwise convergence, 75
of uniform convergence, 75
product, 83
quotient, 71
relative, 62
Sorgenfrey, 102
the coarsest making a given family
of functions continuous, 77
Zariski, 64
totally ordered set, 19
triangle inequality, 24
trigonometric polynomial
complex, 129
real, 129
Tychonoff space, *see* completely
regular space
Tychonoff's theorem, 22, 84, 107,
166
Tychonoff, Andrey N., 107

Uhrysohn's
lemma, 109
metrization theorem, 112
Uhrysohn, Pavel S., 129
uniform continuity, 58
union, 8
universe, 8
upper bound, 19

Weierstraß approximation theorem,
50, 125
constructive proof of, 128
well-ordering principle, 21
Willard, Stephen, 107

Zariski topology, 64
Zermelo–Fraenkel set theory, 21
Zermelo–Fraenkel–Skolem set
theory, *see* Zermelo–Fraenkel
set theory
Zorn's lemma, 19, 107

Universitext   *(continued)*

Jennings: Modern Geometry with Applications
Jones/Morris/Pearson: Abstract Algebra and Famous Impossibilities
Kac/Cheung: Quantum Calculus
Kannan/Krueger: Advanced Analysis
Kelly/Matthews: The Non-Euclidean Hyperbolic Plane
Kostrikin: Introduction to Algebra
Kurzweil/Stellmacher: The Theory of Finite Groups: An Introduction
Luecking/Rubel: Complex Analysis: A Functional Analysis Approach
MacLane/Moerdijk: Sheaves in Geometry and Logic
Marcus: Number Fields
Martinez: An Introduction to Semiclassical and Microlocal Analysis
Matsuki: Introduction to the Mori Program
McCarthy: Introduction to Arithmetical Functions
McCrimmon: A Taste of Jordan Algebras
Meyer: Essential Mathematics for Applied Fields
Mines/Richman/Ruitenburg: A Course in Constructive Algebra
Moise: Introductory Problems Course in Analysis and Topology
Morris: Introduction to Game Theory
Poizat: A Course In Model Theory: An Introduction to Contemporary Mathematical Logic
Polster: A Geometrical Picture Book
Porter/Woods: Extensions and Absolutes of Hausdorff Spaces
Radjavi/Rosenthal: Simultaneous Triangularization
Ramsay/Richtmyer: Introduction to Hyperbolic Geometry
Reisel: Elementary Theory of Metric Spaces
Ribenboim: Classical Theory of Algebraic Numbers
Rickart: Natural Function Algebras
Rotman: Galois Theory
Rubel/Colliander: Entire and Meromorphic Functions
Runde: A Taste of Topology
Sagan: Space-Filling Curves
Samelson: Notes on Lie Algebras
Schiff: Normal Families
Shapiro: Composition Operators and Classical Function Theory
Simonnet: Measures and Probability
Smith: Power Series From a Computational Point of View
Smith/Kahanpää/Kekäläinen/Traves: An Invitation to Algebraic Geometry
Smorynski: Self-Reference and Modal Logic
Stillwell: Geometry of Surfaces
Stroock: An Introduction to the Theory of Large Deviations
Sunder: An Invitation to von Neumann Algebras
Tondeur: Foliations on Riemannian Manifolds
Toth: Finite Möbius Groups, Minimal Immersions of Spheres, and Moduli
Van Brunt: The Calculus of Variations
Wong: Weyl Transforms
Zhang: Matrix Theory: Basic Results and Techniques
Zong: Sphere Packings
Zong: Strange Phenomena in Convex and Discrete Geometry